T0134338

Knowledge Organization in Subject Areas
Vol.1 (KOSA-1)

published by the

International Society for Knowledge Organization

ISSN 0946-9389

First European ISKO Conference
held at SUZA, Facilities of the Ministry of Foreign Affairs
of the Slovak Republic, Bratislava, Slovakia, 14-16 Sept.1994

Organizers:
The National Slovak ISKO Chapter NISKO

in cooperation with
The International Society for Knowledge Organization
Woogstr. 36a, D-60431 Frankfurt/M, Germany

Conference Chair:
Dr.Pavla Stančikova

Program Committee:
Dr. Pavla Stančiková, CEIT, Bratislava (Chair)
Dr. Ingetraut Dahlberg, ISKO, Frankfurt (Co-Chair)
Mr. Christian Galinski, Infoterm, Vienna (Co-Chair)
Dr. Gerhard Budin, Termnet, Vienna (Co-Chair)
Mr. El-Hassane Bendahmane, UNEP-INFOTERRA, Kenya
Mr. Heiner Benking, FAW, Ulm
Dr. Maria Domokos, Reg.Authority
      f. Environm.Protection, Budapest
Dr. Ivan Duša, Ministry of Environment, Bratislava
Prof. Dr. A. Dzhincharadze, VNIIKI, Moscow
Dr. Bruno Felluga, Inst.Technol.Biomedichi, Roma
Prof.Akemi Haruyama, Japan
Mr. Han Heijnen, Int. Water & Sanitation Centre
      Netherlands
Ms. An Hua, Natl. Environm. Protection Agency, China
Ms. Lubica Kosková, Ministry of Foreign Affairs, Bratislava
Dr. Helmut Lessing, Ministry of Environment, Hannover
Dr. Werner Pillmann, CEDAR, Vienna
Ms. Ulla Pinborg, Ministry of Environment, Copenhagen
Prof.Dr. Roland Scholz, ETH Zürich
Dr. Konrad Zirm, Ministry of Environment, Vienna

Environmental Knowledge Organization

and

Information Management

Proceedings
of the
First Euroepan ISKO Conference
14-16 Sept. 1994
Bratislava, Slovakia

organized by the
National Slovak ISKO Chapter NISKO
the ISKO General Secretariat and
Infoterm, Vienna

Edited by

Pavla Stančiková and

Ingetraut Dahlberg

Frankfurt/Main
INDEKS VERLAG
1994

Predocumentation

Stančikova, P., Dahlberg, I.: Environmental Knowledge Organization and Information Management. Proc. 1st European ISKO Conference, 14-16 Sept. 1994, Bratislava, Slovakia. Frankfurt/M: INDEKS Verlag 1994. 224p., ISBN 3-88672-600-2 = Knowledge Organization in Subject Areas, Vol.1 (KOSA-1), ISSN 0946-9389
The volume contains 25 papers and 7 abstracts of papers to be presented at the conference and arranged according to the following 12 topics: Inter- and Transdisciplinary Aspects - Catalogs of Data Sources - National Environmental Information Systems - Classification Systems and Thesauri in the Environmental Sciences - Standardization and Legal Provisions - Conceptual Bases of Environmental Knowledge Organization - Glossaries and Databases in Special Fields - KO in Special Fields - Software and Multilingual Tools - Environmental Education Programs - Environmental Networking on an International and National Scale - Approaches to Harmonize Environmental Information - Future Aspects. - The Keynote Lecture was given by A.J.N. Judge on *"Spherical configuration of categories to reflect systemic patterns of environmental checks and balances"*. Appended is a list of the authors and a name and subject index.

(c)   1994 by INDEKS Verlag, Frankfurt/Main
Umschlaggestaltung: AVIVA W.Dahlberg
Druck: Druckerei Guntrum II, KG, Schlitz/Hessen
Printed in Germany
ISBN 3-88672-600-2; ISSN 0946-9389

Preface

The first European Conference on Environmental Knowledge Organization represents a new commitment of ISKO to tackle KO problems in a subject-oriented field. This field is particularly challenging because of its multidisciplinary character, with its special problems of concept relationships and the need to link these concepts to those from other fields of knowledge. The problems of literature and data documentation are formidable in addition.

Because the environmental sciences are at present one of the most important fields for the future of mankind, ISKO and its conference co-organizers hope that this first endeavour will bear good fruits and that the contents of the papers and the suggestions offered will help to solve the most severe problems in this area and will be useful for all those engaged in this field and seeking advice at present through knowledge organization.

We are grateful that the majority of our speakers were able to deliver their papers early enough so that we could have the conference volume printed in advance. Regarding the few others we decided to include in the volume at least the abstracts of their papers.

The papers are arranged according to the preliminary program. An index provides access also to the names of the authors.

We apologize for any mistakes in the texts still not corrected. We would have liked to publish a more perfectly edited book, but for lack of necessary time this was not possible.

ISKO starts with this volume a new series ("Knowledge Organization in Subject Areas" - KOSA) in order to provide for similar further activities in other areas of knowledge in accordance with the need felt by their representatives.

We herewith thank all those who have encouraged us to organize this conference and who have given support in one way or the other. May this wonderful help find its blessings from above!

Pavla Stančikova   Ingetraut Dahlberg

Bratislava/Frankfurt, Aug.17, 1994

Contents

* only abstract of paper

Environmental Education Programs

Environmental Networking on an International and National Scale

Approaches to Harmonize Environmental information

Anthony J N Judge
Union of International Associations, Brussels, Belgium

Spherical configuration of categories to reflect systemic patterns of environmental checks and balances

Abstract: To explore and illustrate new conceptual and organizational possibilities, the focus of the exercise described is on identifying "strategic dilemmas" underlying debates on environment/development issues, such as those of the 1992 Earth Summit. These are the dilemmas which reflect such seemingly irreconcilable concerns as "safeguarding watercourses" versus "exploiting essential hydro-electric energy reserves". The assumption is that as a set these local (namely issue-specific) long-term dilemmas may offer clues to new patterns of global (namely inter-sectoral) strategies and bargains. In an effort to move beyond the questionable "linearity" of conventional representations of such categories, the information is encoded or projected onto a network derived from a symmetric polyhedral form. The network has been deliberately chosen to facilitate comprehension of global properties of the pattern of strategic dilemmas. However the global significance of the pattern, and the basis for its "sustainability", is shown as emerging only when its form in three-dimensions becomes apparent as having spherical characteristics. Of special interest is the shift in the level of analysis from isolated problems to that of vicious cycles or loops that link a succession of problems aggravating one another across conventional categories. The concern is how these cycles interlock, defining such a sphere, in order to sustain negative environments which call for transformation. The research is based on data prepared for the 4th edition of the *Encyclopedia of World Problems and Human Potential* (1).

1. Introduction

There is an increasing sense of urgency in the international community. This is accompanied by an increasing sense of opportunity. The urgency relates to the perception of an ever-increasing pressure from world problems and their effects at every level of society. The opportunity is associated with the many tools, insights and resources available to society, especially in the field of information.

It is easy to get caught up in the enthusiasm and hype concerning the future "information society", the "global village", and the "superhighway". It should not be forgotten that the automobile excited the same enthusiasms at the beginning of the century but led to unforeseen impacts on the environment, on quality of life and on the marginalization of those without access to such advantages. There will undoubtedly be advantages and many are already accessible to the few. But what will be the equivalents of the proliferation of roads, noise, pollution, conflicts between public and private transport, unaesthetic roadside "furniture", road accidents, and the like? The information revolution will engender its own "environmental" problems as can already be seen in the invasive spread of commercial and political messages.

Even vaster quantities of information are about to become widely available. It would be a grave mistake to assume that this in itself will necessarily empower people and groups to respond more effectively to the social and environmental problems of the immediate future. The enthusiasms of many information professionals and providers tend to conceal the inadequacies of the existing tools and those which are on the drawing board.

2

The key challenge is whether the tools will empower people to ask more appropriate questions or whether they will simply reinforce users in the pursuit of predetermined preoccupations. Will users emerge with unforeseen answers that offer them a more integrative perspective, or will mindsets that are already endangering society simply be reinforced? There should be no illusions about the marketing of information tools in a society that will be encouraged towards the highest level of information consumption. As ever, the money is to be made in providing people with responses to their immediate, short-term needs not in challenging them to reframe their needs and their approach to information about them.

The above remarks apply as much to individual end-users as to major institutions, notably those operating at the international level.

2. Environmental information: the neglected challenge
It is fair to state that most effort in connection with environmental information is concerned with the more immediate practical tasks of obtaining that information (including environmental monitoring), and incorporating it into information systems of various levels of technical sophistication in such a way that it can be readily retrieved. It is also fair to state that, in organizing this information, practically all effort focuses on the traditional bibliographic elements of author, title, publisher and subject, together with whatever investment is appropriate in abstracting, thesaurus design and enabling users to formulate complex search profiles, possibly across a complex set of dispersed databases. For many information professionals, if this alone was achieved, their responsibility could be considered honourably discharged. For some there is the supplementary challenge of making the information widely available (notably across language and script barriers), especially to those who are already marginalized by the exorbitant costs of many of the better information tools. These issues blur into the new multi-media opportunities of disseminating information.

The question to be asked is whether information tools are being effectively designed to empower people to comprehend the nature and problems of complex systems and the windows of opportunity associated with them. Is it possible that preoccupation with "user friendliness" may be obscuring fundamental conceptual challenges and their organizational implications?

This question goes beyond issues of storage and retrieval. It also goes beyond preoccupation with the many flashy features of multi-media access to large dispersed data bases. There is a need to overcome the tendency for the much acclaimed advances in storage, retrieval and presentation techniques to disguise what is not being achieved in enabling improved comprehension and decision-making with respect to increasingly complex systems -- namely the failure to address fundamental conceptual issues of knowledge organization. This is especially the case where policy-making is fraught with intractable differences, dilemmas and paradoxes, as is increasingly apparent at the highest level -- notably with respect to environment and development issues. There is a need for information systems to take explicit account of support, opposition or incompatibility between documents rather than simply covertly excluding those deemed irrelevant or inappropriate.

3. Background
The remainder of this paper builds on aspects of a long-term programme of the Union of International Associations (Brussels) which results periodically in the publication of an *Encyclopedia of World Problems and Human Potential*(1). This is a companion volume to the better known *Yearbook of International Organizations* (2). Both publications are generated from large databases on some 15,000 international organizations (governmental and nongovernmental) and on some 9,800 "world problems" with which they are preoccupied. Such information raises many issues of knowledge organization if it is to be of greater value

than a telephone directory.

The UIA has had a long tradition of concern for organization of information, dating from its founders involvement in the UDC (prior to the creation of the FID) through to involvement as rapporteur for the two international symposia on United Nations documentation (3, 4) and the early work of COCTA (5). The above-mentioned Encyclopedia has, since its first edition in 1976, focused on the integrative and interdisciplinary challenges of knowledge organization. The functional classification of organizations in the Yearbook has also been a preoccupation (6). These concerns have led to involvement in the debate on information relevant to new forms of governance (7) and to the organization of the forthcoming 1st World Congress on Transdisciplinarity (Portugal, November 1994).

The specific challenge of environmental and developmental information was confronted most recently in a background document (8) prepared for an Inter-Sectoral Dialogue on the occasion of the Earth Summit (Rio de Janeiro, 1992). This endeavoured to indicate possibilities for a new kind of "global" framework to interrelate the many "local" (ie specialized) issues identified in Agenda 21 and related documents. The bias in what follows is towards elucidating, in an experimental spirit, the nature of such a global framework on the assumption that users with specialized needs tend to be well-served by current information systems. In this sense the strategic challenges for the future are at the "global" level.

4. Global knowledge organization

"Global" is used here to imply the full world-wide range of environmental and developmental issues in the broadest sense, as well as the manner in which that knowledge is held for comprehension as a whole -- especially in relation to democratic policy-making. "Knowledge" is seen as necessarily transcending the preoccupations of single disciplines -- especially when this involves strategic dilemmas and paradoxes that emerge between disciplines confronted with the need to act on problems of the environment. Under such circumstances it is assumed that "organization" will necessarily involve non-linear dimensions which may appear incompatible with conventional approaches to knowledge organization. The challenge is sadly illustrated by the marked reluctance to organize comprehensive species databases to enable complex food webs to be effectively mapped, notably to help rehabilitate degraded environments.

Given the intractable policy differences in relation to environmental and developmental issues, there is a strong case for exploring a form of global organization which makes structural use of such differences. The common assumption that they can be ignored (as being purely the concern of end-users), results in information systems which do nothing to address the issue (faced by key end-users) of interrelating incommensurable policy perspectives. The resulting design is then only of marginal value to those who are obliged, through their strategic responsibilities, to deal with such incommensurability.

The possibility explored here is the need to move beyond preoccupation with "descriptors". These may well be absolutely necessary to access information describing specific issues. But the conceptual challenge of environmental issues lies primarily in the way that clusters of information challenge each others validity in any global context. The significance, scope, stability and relevance of descriptors is then severely challenged. Information professionals must either learn to deal with this challenge or be by-passed by other kinds of system (however crude) that will meet these more complex needs of policy-makers. This is the realm of strategic dilemmas and paradoxes in which the significance of categories is continuously subject to flux. For example, oestrogenic effects may soon completely reframe environmental issues and priorities.

5. Configuring "global bargains" through more complex structure

In preparation for the 1992 UNCED event, as a follow-up report to his involvement as Secretary-General of the World Commission on Environment and Development (responsible for the Brundtland Report), Jim MacNeill (9) articulated for the Trilateral Commission the policy options for sustainable development in terms of "**shaping global bargains**". He notes: *"The notion of a 'global bargain' conjures up many images, especially within the broad context of sustainable development...In its simplest terms, a bargain involves at least two parties and two issues. It implies a trade-off between the parties on the issues. The group of nations, developed and developing, that have come together to form a bargain must agree to give up something in order to get something else. As a rule, they would give up a path of development in a given sector that is unsustainable and thus represents a threat to themselves and the other negotiating nations or the global commons."*

According to this perspective the arenas to be subject to bargaining emerge haphazardly as a result of conventional political processes. There is no systemic sense of how the bargains interweave to ensure the sustainability of development as whole. There is no sensitivity to issues which can be conveniently ignored by powerful majorities. In a real sense this corresponds to the traditional paradigm of ad-hocery which has contributed so much to the emergence and maintenance of the present crisis. Note that current information systems do little to compensate for this ad-hocery which they so effectively reinforce.

The difficulty is that bargains are typically discussed in the verbal and textual mode. In this mode, notions of "giving up" in order to "get something else" are understood in the simplest terms and therefore readily evoke opposition. This opposition is indeed legitimate in terms of the "two-dimensional" images (of "sides") through which they are currently discussed. It would not however be so necessary in terms of more complex configurations (of "sides").

6. Beyond isolated bargains

It is unfortunate, as the MacNeill report illustrates, that thinking for the 1992 Earth Summit was focused on the possibility of a series of issue-specific "global bargains". Taken one by one, these may or may not prove negotiable or effective. But on this basis there is every likelihood that the effects of some will significantly undermine the effects of others. What is missing is any image of how issue-specific bargains can be interwoven to constitute a larger sustainable development bargain -- as a set of complementary elements rather than as an asystemic jumble.

As in architecture, it is through balancing the stresses and tensions between a set of complementary construction elements that the integrity of a building is ensured. Richer structured imagery is required to facilitate understanding of how the larger and more encompassing bargains can be achieved. It is through such images of integrity, emerging from more complex structures, that the logic of that integrity gives justification to issue-specific bargains of greater effectiveness. It shows how they "fit". Structured images are required to give precision to the vague notions of "checks and balances" conventionally articulated in textual terms. Such images give precision to the notions of "giving up", and tensional "trade offs", which readily lend themselves to description in architectural terms, for example.

The overall purpose of any inter-sectoral dialogue is to raise the level of inter-sectoral debate. The challenge is to move beyond simplistic consensus and beyond acrimonious restatement of established positions. The challenge is to move towards "higher orders of consensus".

7. Strategic dilemmas

The exercise undertaken for the Inter-Sectoral Dialogue focused on identifying "strategic dilemmas" underlying debates on Earth Summit issues. These are the dilemmas which reflect such seemingly irreconcilable concerns as "safeguarding watercourses" versus "exploiting essential hydro-electric energy reserves". The assumption here is that the set of these local (namely issue-specific) long-term dilemmas may offer clues to new patterns of global (namely inter-sectoral) strategies and bargains.

The starting points were a brainstorming exercise in the identification of **polarizing dilemmas** and the clustering of some 450 issues identified in the Brundtland Report, Agenda 21, and in sectoral declarations (8). As a checklist the latter had the merit of providing a crude context for specific sectoral concerns. However this was not enough. It failed to respond to the need to raise the level of debate by offering a global (inter-sectoral) context for specific bargains, checks and balances. Such checklists, like Agenda 21, are effectively overwhelming. They encourage simplistic attempts to identify "the most important problem" whose solution it is hoped will magically transform all the others.

8. Pattern of strategic dilemmas

Figure 1 (a and b) is one attempt to respond to this situation by showing how different social functions, understood as strategic opportunities, interfere with each other to engender a pattern of strategic dilemmas. In that pattern each strategy may take a privileged role or may in turn be constrained by other strategies. For example, when "environment" is a privileged function, "well-being (+jobs)" may be sacrificed, whereas, when "well-being (+jobs)" is the privileged function, sacrificing "environment" is the alternative option. Neither option is satisfactory, but both appear to have their place. Any such dilemma may of course be "resolved" by short-term measures, but the nature of the dilemma renders such solutions unsustainable in the longer-term. **Sustainable development is a function of the pattern as a whole rather than of its components.**

The choice of six principal functions as the basis for the pattern in Figure 1 is of course arbitrary -- but it is certainly more systemic than the chapter organization of the Brundtland Report or of Agenda 21. As noted below, a different number of clusters could have been used, bearing in mind the constraints of over-simplification and excessive complexity.

9. Network of bargain arenas

The traditional tabular presentation of Figure 1 is itself a conceptual trap. It encourages a very mechanistic approach to the pattern of dilemmas, reinforcing tendencies to much-contested forms of "linear thinking". The linearity may be deliberately challenged by allowing the information to be encoded or projected onto a network. In the light of the arguments elsewhere concerning polyhedral nets (1, Section TZ; 7, 10), in this exercise **the network was deliberately chosen to facilitate comprehension of global properties of the pattern of strategic dilemmas** by mapping the information in Figure 1 onto an icosadodecahedral net (see Figures 2A and 2B). As noted below, the global significance of the pattern, and the basis for its sustainability, only emerges when its form in three-dimensions become apparent.

In the network the principal lines traversing the pattern are used to represent the six selected strategic preoccupations of Figure 1. They are coded by the same letters. In two dimensions most of the relationship lines can only appear as broken, although in three dimensions they are seen to form unbroken interlocking circles defining the integrity of a sphere, as is seen when the original polyhedron is reconstructed in 3 dimensions (see Figures 4 and 5). In this exercise, the interlocking of these circles creates a pattern of triangles and pentagons on the surface of the sphere (more apparent in Figure 5). These may be understood as simpler (3-valent) and more complex (5-valent) bargaining arenas around specific concerns.

10. Identifying the bargaining arenas (as "strategic categories")

Each triangle in the network can be described by a 3-letter code reflecting a particular combination of the original 6 strategic functions. On the basis of work on coding the declaration issues according to these functions, a tentative indication of the significance of each code was formulated in two versions (see Figures 2 and 3): one indicated a **development-focused application of the strategies**; a second indicated **an environment-sensitive application of the strategies**. In both cases typical problems resulting from inappropriate implementation could be identified. Keywords from that indication have been inserted into the network diagram.

It becomes clear that on a single network pattern (Figure 2A), two triangles appear with th same code, and are therefore used here to indicate the development-focused and the environment-sensitive keywords for that code combination. They are on opposite sides of the network (notably when displayed in three dimensions). Only half of the 20 possible combinations appear on that pattern. A further 10 appear in the second version (Figure 2B). The two versions result from the different orders in which the functions can appear. The full range of Earth Summit issues and strategies is effectively mapped onto these two networks.

It is necessary to use two alternate versions of the network pattern with this approach. This may not be the case with other coding approaches along these lines. Complementary projections are however also required in geographical mapping. Organic molecules essential to life (notably benzene) are based on resonance between two complementary structures. Most tensegrity structures (see below) exist in right- and left-handed versions.

11. Re-interpreting the global bargaining challenge
In contrast to the Earth Summit approach, **the patterning exercise here emphasizes the necessarily global structure of the network of issue-specific bargains**. Namely it starts from an assumption of inter-sectoriality (functional globality) and allows specific sectoral (functionally local) concerns to emerge as features of the pattern of strategic options. From this perspective, **it seems extremely doubtful that local issue-specific bargains (emissions, forests, etc) can be effectively struck in isolation from the global context of strategic dilemmas** -- as tends currently to be assumed. Any such isolated bargains would therefore tend to be unsustainable in the longer-term.

This perspective does however suggest that articulation of these dilemmas within a global framework may redistribute the tensions which currently make it extremely difficult to achieve issue-specific bargains of any consequence in isolation. This redistribution may well provide unsuspected contextual support for such bargains by rendering explicit a new pattern of checks and balances. **Where bargains are no longer treated in isolation, tensions which would otherwise have to be dealt with explicitly within a given bargaining arena (reducing the probability of success) may now be recognized implicitly as contextual to that bargain.** This stresses the importance of treating the totality of Earth Summit issues as a set of inter-weaving strategic options in order to reduce the difficulty of achieving success on particular fronts. This is a specific challenge to the design of information systems and interfaces.

This approach **points to new policy possibilities in which the degree of global consensus required is reduced to a minimum** (in a design sense) by localizing the patterns of disagreement. In this way **disagreement no longer acts globally -- tearing apart the global community.** Rather it is locally confined and understood as a long-term strategic dilemma on which "consensus" can only be achieved in the short-term. **Sustainability thus lies at the global level not at the local level.**

12. Catalytic imagery
There is a need for richer, and more challenging, imagery to capture the complexity of

strategic options to clarify new options both for policy makers and wider audiences. The two-dimensional representation, for "local" purposes, of the "global" structure of the Earth clarifies the challenge. The importance of the shift to three-dimensional representation is particulary obvious in this geographical parallel between representations of the Earth as a globe, and the many efforts to project such information onto 2-dimensional maps -- each with their special distortion. It is the inadequacy of the 2-dimensional representation which highlights the value of the 3-dimensional structure in stressing **globality** and providing a context for **local** issue-specific arenas.

Both in the two- and three-dimensional forms the imagery proposed here is **an invitation to reflection along new lines**. As intended, it deliberately breaks with familiar patterns. It invites further reflection and experiment to better portray the relation between global and local -- and the strategic opportunities which emerge. It is possible that the main value of the structures presented lies in the mapping exercises that they encourage, namely in the creativity and reflection that they evoke, rather than in any particular pattern which may be favoured. A "structural outliner" has been proposed (11) as a specific software package to facilitate such exploration by end-users, notably of transformations between the structures in Figure 4. "Packing" and "unpacking" systems of categories, according to needs for detail or simplicity, is one of the advantages of using the system of interrelated polyhedral forms of Figure 4.

13. Possible interpretation refinements
The merit of the 3-dimensional representation of the Earth Summit issues is that it may be used to clarify why strategic dilemmas appear to emerge. Bargain arenas have been recognized here in pairs of triangles in a network pattern. The **"dilemma" in each case may be seen as a failure to recognize the global properties of the structure** which separate the two complementary (but distinct) arenas -- for these are on opposite sides of the spherical structure. Collapsing the distinctions into a two-dimensional representation, in which the triangles are super-imposed, is what guarantees the appearance of a dilemma. It is an appropriate global consensus which allows them to be understood as separate, thus eliminating the dilemma.

In practice the construction of three-dimensional spherical structures (like geodesic domes) requires understanding of more than those surface features with which the bargaining arenas have been associated here. According to the principles of tensegrity (namely tensional integrity) explored by R Buckminster Fuller (10), new types of self-sustaining global structure may be created by a particular three-way pattern of tensile forces. Such a structure is not supported or maintained (by special authority structures). It is pulled outward into sphericity by inherent tensional forces which its geometry also serves to restrain (10). It responds as a system with local stresses being uniformly distributed throughout the structure, and uniformly absorbed by every part of it as a classic example of synergy. It is not necessary that these structures should be patterned on regular polyhedra, but the tension networks are most economical when their strands run for considerable distances without changing direction -- and preferably along great-circles. As the work of cybernetician Stafford Beer (12) is illustrating, there is every possibility that such structures may come into their own to ensure the sustainability and integrity of Internet conferences.

Tensegrity structures clarify ways in which individual bargains need to be interlocked using local elements of disagreement ("compression elements") within the global network of agreement ("tension elements"). **Tensegrity structures are effectively patterns of sustainability.** The challenge is to find useful ways to encode such patterns to offer insights into the strategies of sustainable development.

It is important to recognize that there are whole families of network patterns that correspond to different spherical structures in three dimensions (see Figure 4). Figures 2 and 3 suggest just one way of "cutting" the conceptual or strategic "cake". There are indications that increasing the complexity of the network in order to explicitly capture more detailed issues could provide global contexts which make it even easier to handle issue-specific bargains. What is required is a special database which could enable people to shift between different levels of functional detail as is done between maps in geographical atlases and in geographical information systems.

14. Vicious cycles and loops

The arguments above stress the systemic significance, at the global level, of interlocking circular sequence of relationships. To take the investigation further data is required to clarify the nature of such relationships in practice and across a wide spectrum of issues of relevance to environment and development in the broadest sense.

There has long been recognition of how one problem can aggravate another and of how several problems can reinforce each other. Volume 1 of the *Encyclopedia of World Problems and Human Potential* (1) registers some 120.000 relationships between 9,800 problems in complex networks. Much of the information derives from documents of international organizations. Clearly such relationships may form chains or pathways linking many problems. But hidden in the data as presented is also the existence of chains that loop back on themselves, especially chains of problems that aggravate one another in succession. The more obvious loops may be composed of only 3 or 4 problems (see Figure 6). Far less obvious are those composed of 7 or more. An example is: *Alienation > Youth gangs > Neighbourhood control by criminals > Psychological stress of urban environment > Substance abuse > Family breakdown > Alienation.* No systematic attempt seems to have been made to identify such vicious relationship loops or cycles through which four or more problems constantly reinforce one another. Such cycles are vicious precisely because they are self-sustaining.

A computer program has been developed as an experiment to explore the many pathways amongst the world problems documented in the Encyclopedia database in order to isolate such loops. This suggests the possibility of moving from a focus on problems as though they were isolated, of which few are, to one in which the focus is on the many vicious loops of which a problem may be a member.

15. Cycles as a unit of analysis

Functionally and conceptually such vicious cycles may offer a better way to approach complex networks of problems. Indeed they serve to make clear that any organization with projects focusing on a single problem needs to be aware of any vicious cycle of which that problem is a part. Unless that organization coordinates its activities with any bodies focusing on other problems in that cycle, its work may be totally undermined. Despite apparent success in responding to a particular problem in the short term, this may not affect the sustainability of the vicious cycle of which it is a part. The problem may be regenerated by pressures building up in other parts of the cycle.

It is interesting to reflect on the parallel to the metabolic pathways documented by biochemists. Analysis of such pathways has shown the presence of a number of cycles. Many of these now bear names (eg the glyoxylate cycle, triglyceride pathway) because of their literally vital importance to life processes. There is every possibility that cycles linking problems could prove of equivalent importance -- and possibly of greater strategic significance than the individual problems themselves, since it is the cycles that effectively sustain the individual problems.

5. Strategic responses to problem cycles

Ideally a coalition of organizations should form in response to each vicious problem cycle. Information passed between organizations should then match the pattern of impacts between the problems with which they are concerned. The strategic issue may be less one of how to "break" the cycle and more one of how to reverse it, exploiting the fact that problems are functionally linked in this way.

A significant number of problems also alleviate other problems, although this information is less easy to obtain and such links are consequently less frequent in the Encyclopedia. There may therefore be beneficent problem cycles through which problems constrain each other. It could prove strategically advantageous to locate such cycles.

16. Reservations, results and examples

Before commenting on the experiment in detecting vicious cycles, it is important to recognize that it is precisely through the detection of such loops that attention can be drawn to defects in the pattern of relationships in the data. Detection of loops is therefore in the first place an editorial tool. It raises questions as to the appropriateness of certain links which otherwise may go unquestioned. It also sharpens the discussion on how distinctions are made, using verbal categories and definitions, and how system boundaries are drawn grouping what is represented in this way. Because of the priority given to revising the pattern of relationships there was insufficient time to run the program in anything but a test mode. It was not possible to correct obvious defects before running it to detect substantively significant loops only. Nevertheless the results are indicative of a very interesting area for further exploration.

Using several 386 and 486 machines in parallel, some 9,519,722 pathways were tested for loops involving up to 7 problems. This process identified 7,303 such loops (3Loop=35; 4Loop=115; 5Loop=527; 6Loop=3,058; 7Loop=3,568). The procedure needs refinement, notably to detect problems that are not in loops.

17. Configuring interlocking cycles

Given the possibility of identifying such cycles, the question raised by the earlier discussion is how this information could be best portrayed through various mapping techniques. One attractive possibility, consistent with the spherical emphasis above, is to map the circles around the surface of a sphere with whatever interlocking the data implies. More sophisticated software is required to "massage" such circles around the spherical forms in approximation to structures such as those in figure 5.

The challenge of global governance is to match the complexity of the resulting structure by communication pathways (notably through Internet conferences) and organized initiatives.

References

(1) Union of International Associations. Encyclopedia of World Problems and Human Potential. München, K G Saur Verlag, 4th ed., 1994, 3 vols.

(2) Union of International Associations. Yearbook of International Organizations. München, K G Saur Verlag, 1994, 31st ed, 3 vols.

(3) Judge, Anthony J. N.: Acquisition and organization of international documentation. In: International Federation for Documentation: Proceedings of the International Symposium on the Documentation of the United Nations, Geneva, August 1972. Den Haag, FID, 1974, FID publ. 506, pp 112-144 (UNITAR/EUR/SEM1/WPII/IR)

(4) Judge, Anthony J. N.: Utilisation of international documentation. In: Dimitrov, Th. (Ed.) International Documents for the 80's; their role and use. Proceedings of the Second World Symposium '

10

on International Documentation (Brussels, 1980). New York, Uniflo, 1982.

(5) Judge, Anthony J. N.: Anti-developmental biases in thesaurus design. In: Riggs, F. W. (Ed.): The CONTA Conference: proceedings. Frankfurt/Main, Indeks Verlag, 1981, pp 185-201

(6) Judge, Anthony J. N.: Functional classification. International Classification 11 (1984) 2, pp 69-76; 3, pp 139-150.

(7) Judge, Anthony J. N.: Guiding metaphors and configuring choices. 1991 (Paper for the Development Administration Division of the United Nations Department of Technical cooperation for Development (UN/DAD/DTCD) to appear in a book on "Tools for Critical choice by Top Decision Makers")

(8) Judge, Anthony J. N.: Configuring globally and contending locally; shaping the global network of local bargains by decoding and mapping Earth Summit inter-sectoral issues. Brussels, Union of International Associations, 1992 (Prepared for the International Facilitating Committee for the Independent Sectors in the UNCED process)

(9) MacNeill, Jim et al. Beyond interdependence; the meshing of the world's economy and the world's ecology. New York, Oxford University Press, 1991 (A Trilateral Commission book)

(10) R Buckminster Fuller. Synergetics; explorations in the geometry of thinking. New York, Collier, 2 vols, 1975-1982

(11) Judge, Anthony J. N.: Catalyzation of new patterns of collaboration using a PC-based structural outliner as an imaging scaffold. Brussels, Union of International Associations, 1992

(12) Stafford Beer. World in torment; a time whose idea must come (Presidential address to Triennial Conference of the World Organization of Systems and Cybernetics, New Delhi, 1993). Includes summary of syntegrity scheduled to appear in book form under the title: Beyond dispute; the invention of team syntegrity.

INTER-SECTORAL STRATEGIC DILEMMAS OF SUSTAINABLE DEVELOPMENT

Constrained function \ Privileged function	Population Security — P	Well-being Health — W	Learning Education — L	Production Trade — T	Environment Impacts — E	Regulation Equity — R	UN Bodies	Sectors
Population/Relief, Security/Peace, Vulnerable groups, Women/Youth — **P**	Sacrifice of one population group for another? PP	Sacrifice pop. relief/growth for well-being of population WP	Sacrifice pop. relief/growth for education, research, etc LP	Sacrifice pop. relief/growth for economic growth? TP	Sacrifice pop. relief/growth for environment? EP	Sacrifice population relief/growth for equity? RP	UNFPA, UNV Security CI UNICEF, UNHCR INSTRAW	Relief Military,Peace Indigenous Women,Youth
Well-being/Health, Employment (condit.), Quality of life, Welfare/Fulfilment — **W**	Sacrifice well-being (+jobs) for population relief/growth PW	Sacrifice of one form of well-being for another? WW	Sacrifice well-being (+jobs) for education, research, etc LW	Sacrifice well-being (+jobs) for economic growth? TW	Sacrifice well-being (+jobs) for environment? EW	Sacrifice well-being (+jobs) for equity RW	WHO ILO HABITAT,UNRI SD	Health Labour Religions
Learning/Education, Science/Research, Culture/Arts/Lang., Inform./Communic. — **L**	Sacrifice educ. (+culture) for pop. relief/ growth? PL	Sacrifice educ. (+culture) for well-being? WL	Sacrifice of one form of education for another? LL	Sacrifice educ. (+culture) for economic growth? TL	Sacrifice educ. (+culture) for environment? EL	Sacrifice educ. (+culture) for equity? RL	UNESCO, WIPO UNU, UNITAR UPU,ITU,ACCIS	Students Science Media Information
Production/Trade, Industry/Technology, Agriculture/Fish, Energy/Mining — **T**	Sacrifice prod. (+trade) for population relief/growth PT	Sacrifice prod. (+trade) for well-being? WT	Sacrifice prod. (+trade) for education, research? LT	Sacrifice of one form of production for another? TT	Sacrifice of prod. (+trade) for environment? ET	Sacrifice of prod. (+trade) for equity? RT	UNDP, UNCTAD UNIDO,GATT FAO,WFP,IFAD IAEA	Development Industry/ Commerce
Environment/Impact, Ecosystems/Species, Urban / Transport, Design/Landscaping — **E**	Sacrifice envir. for population relief/growth? PE	Sacrifice envir. for well-being? WE	Sacrifice envir. for education, research, etc? LE	Sacrifice envir. for economic growth? TE	Sacrifice of one environmental benefit for another? EE	Sacrifice envir. for equity? RE	UNEP, UNESCO HABITAT, WMO ICAO, IMO	Environment Conservation Architects
Regulation/Equity, Govern./Mgt./Admin., Justice/Order, Finance/Debt Mgt. — **R**	Sacrifice equity for population relief/growth? PR	Sacrifice equity for well-being? WR	Sacrifice equity for education, research, etc? LR	Sacrifice equity for economic growth? TR	Sacrifice equity for environment? ER	Sacrifice of one form of equity for another? RR	ECOSOC, ACC UNDP ICJ IBRD,IMF	NGO coalit. Law, Rights Finance/Banks

- ◆ Column and row headings correspond to major clusters of functions essential to the sustainable development of society.
- ◆ Words used to label the scope of clusters are necessarily inadequate at the level of generality indicated.
- ◆ The clusters could be "unpacked" to give more columns and rows. Cluster label words would then become more specifically appropriate.
- ◆ The cells of the table indicate fundamental dilemmas of sustainable development.
- ◆ Sacrifices must be made in the process of change and transformation. The dilemma is the level of the sacrifice that is appropriate.
- ◆ Sustainability is determined by the balance between the various forms of sacrifice and the constraints that they impose on one another.
- ◆ The columns/rows can be thought of in terms of values, logics or languages which successively confront each other.
- ◆ Sectors in right hand column are mainly those represented at the Rio de Janeiro Inter-Sectoral Dialogue

Figure 1B

SIGNIFICANCE OF CODES IN FIGURE 1A (tentative)

On the basis of 6 principal functions essential to sustainable development, there are 30 strategic dilemmas which may be grouped as 15 pairs. The indicative texts below may stress specific examples.

PW **Population needs/satisfaction / Social (un)development**
PW: Application of austerity measures to ensure long-term viability of population (cf "structural adjustment")
WP: Avoidance of measures of restraint to satisfy short-term popular demand

PL **Societal culture / Tradition**
PL: Commitment to family and group advancement at the expense of individual education (cf traditional parental commitment, socialist educational priorities)
LP: Commitment to individual education at the expense of family and group advancement (cf "selfish self-advancement". learning without social obligations)

PT **Economic (in)security of social groups**
PT: Foregoing economic opportunities to safeguard cultural integrity (cf indigenous groups, isolationism. restricted employment of women)
TP: Pursuit of economic opportunities despite the effects on cultural integrity and minority groups (cf discrimination in employment, slavery, "sweat shops", child labour)

PE **Environmental (in)security**
PE: Exploitation of non-renewable resources to ensure group survival (cf degradation of marginal lands, deforestation for fuel-wood)
EP: Control of population growth and activity to conserve natural resources

LR **Intellectual disciplines / Systems analysis**
LR: Excellence at price of general improvement in learning
RL: General improvement in learning at the price of excellence

TR **Regulation of trade / Finance**
TR: Rewarding the entrepreneur at the price of the worker (capitalism)
RT: Equal sharing of benefits to the detriment of the most productive (socialism)

ER	**Environmental regulation** ER: Limiting environmental benefits to the privileged RE: Allowing degradation of the environment through unconstrained access to resources
PR	**Social (in)justice / Governance / Law / Power** PR: Allowing one group to infringe upon the rights of another RP: Limiting population expansionism to safeguard vulnerable groups
WR	**Ethical/Moral/Spiritual living** RW: Limiting quality of life (+ jobs) to the privileged WR: Reducing quality of life (+ full employment of the few) so that all benefit, to however limited a degree
WE	**Quality of life** WE: Over-consumption and exploitation of non-renewable resources EW: Reduction in quality of life (+ jobs) to safeguard environment
TE	**Economic development** TE: Over-exploitation of non-renewable natural resources ET: Reduction in industrial and agricultural (+ fisheries) productivity to safeguard natural resources
LE	**Ecology** LE: Undertaking research and traditional cultural activities destructive of the environment (cf "scientific whaling") EL: Foregoing research and traditional cultural activities destructive of the environment
LT	**Research and development / Technology development** LT: Restraining economic development to permit learning (cf ecosystem research, urban archaeological sites) TL: Limiting education to training relevant to employment opportunities (cf "educational relevance")
WT	**Livelihood / Conditions of work / Consumption** WT: Foregoing economic opportunities to cultivate quality of life (indigenous cultures, "dropping out") TW: Economic development at price of health and quality of life ("entering the rat race", "no gain without pain")
WL	**Arts / Cultural self-knowledge** WL: Foregoing further learning opportunities to enjoy present quality of life (cf avoiding challenges) LW: Giving up present quality of life to focus on challenges of further learning education)

Figure 2A

REPRESENTATION OF ISSUE ARENAS ON ICOSADODECAHEDRAL NET (tentative)
(Alternate A)

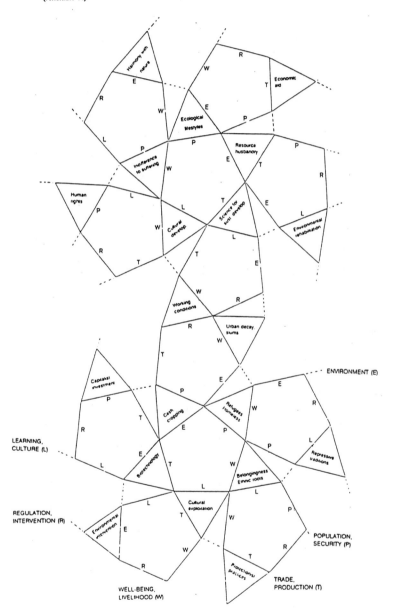

Figure 2B

REPRESENTATION OF ISSUE ARENAS ON ICOSADODECAHEDRAL NET (tentative)
(Alternate B)

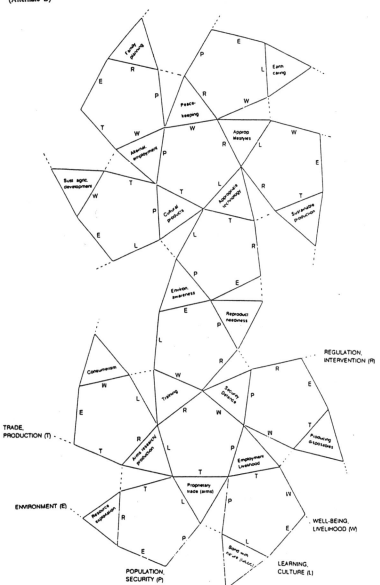

SIGNIFICANCE OF 3-LETTER CODES IN FIGURE 2

Based on combinations of functions presented in Figure 1. Wordings are indicative only. See also Figure 1A.

DEVELOPMENT-FOCUSED

LER	+	Application of knowledge to redesign the environment (eg draining swamps, clearing forests, introduction of species)
	–	Irresponsible intervention in ecosystems (eg elimination of wetlands)
LTE	+	Application of learning to productive exploitation of the environment ("green revolution") / Biotechnology management
	–	Environmentally irresponsible research / Irresponsible biotechnology / Unconstrained advocacy of agrochemicals
LTR	+	Application of knowledge/know-how according to production/commercial priorities (arms production)
	–	Misapplication of knowledge (weapons research)
PER	+	Fulfillment of reproductive needs / Unconstrained population dynamics
	–	Excessive population levels (for the environment) / Unwanted children
PLE	+	Traditional (spiritual) bond with the environment
	–	Taboo / Superstition
PLR	+	Maintenance of traditional patterns of authority (or cultural values) / Imposition of particular ideology
	–	Social rigidity / Discrimination against minorities / Repression / Socially inappropriate conceptual inadequacies
PLT	+	Restrictive proprietary trade (seed varieties) / Arms trade
	–	Dependence on high yield crops
PTE	+	Economic exploitation of environmental endowments
	–	Military/Industrial complex / Cash crops
PTR	+	Capital investment ideology for economic gain (eg capitalism)
	–	Exploitative political/economic ideologies (eg colonialism, fascism)
PWE	+	Environmentally-dependent indigenous communities / Nomadic lifestyles
	–	Unsustainable land/water use / Refugees / Homelessness / Migration / Squatters

ENVIRONMENT-SENSITIVE

LER*	+	Application of knowledge to reconstitute devastated ecosystems
	–	Ineffectual application of knowledge to remedy negative environmental conditions
LTE*	+	Science for sustainable development
	–	Ineffectual use of learning to ensure environmentally sound development
LTR*	+	Appropriate research for economic development / Research into non-exploitive patterns of production
	–	Ineffectual application of knowledge for development
PER*	+	Population control / Family planning
	–	Eugenics / In-breeding / Ageing population / Imbalanced sex ratio
PLE*	+	Environmental awareness
	–	Environmental fascism
PLR*	+	Human rights / "Affirmative action"
	–	?
PLT*	+	Development of culturally sensitive products / Exchange of cultural artefacts
	–	Culturally insensitive development
PTE*	+	Husbandry/Stewardship of available resources
	–	Community collapse / Economic decay
PTR*	+	Economic development programmes / Aid for development / Aid "with human rights strings"
	–	Ineffectual remedial ideologies (eg socialism, communism)
PWE*	+	Creation of ecological settlements (cities/villages) / lifestyles
	–	Socially unattractive/alienating ecological settlements and lifestyles

PWL	+	Belonginess / Roots
	−	Ethnic disintegration / Cultural invasion
PWR	+	Protection of the population / Security / Defence
	−	Socio-political insecurity / Disempowerment
PWT	+	Productive employment / Livelihood security
	−	Unemployment / Food shortages
TER	+	Exploitation of resources for economic development (survival)
	−	Wasting non-renewable resources / Laying waste / Wastes
WER	+	Exploitation of environment to enhance quality of life (eg safaris, all terrain vehicles, exotic plants/pets/cosmetics/foodstuffs/drugs)
	−	Urban decay / Slums
WLE	+	Exploitation of environmental and cultural resources to maximize well-being / Consumerism
	−	Over-consumption patterns
WLR	+	Human resource development / Training for jobs / Educating public opinion / Public health programs
	−	Inappropriate education / Propaganda
WLT	+	Exploitation of cultural heritage / Exploitation of commercial opportunities of tourism
	−	Degradation of cultures / Cultural invasion
WTE	+	Energy-intensive production to improve quality of life / Exploitation of energy resources for development (survival) / Production of non-essentials (throw-away products)
	−	Over-exploitation of environmental resources (eg over-fishing, over-grazing, fuel-wood crisis, "slash and burn")
WTR	+	Protection of trade, production and traditional livelihoods
	−	Restrictive trade practices / Exploitative conditions of work / Crime

PWL*	+	Charitable action
	−	Indifference to condition/suffering of others
PWR*	+	Peace-keeping operations / Empowering communities
	−	Miscarriage of justice / Police brutality
PWT*	+	Development of alternative employment opportunities
	−	?
TER*	+	Sustainable use of resources / Restraint on exploitative growth
	−	Ineffectual/Unproductive use of resources (for survival)
WER*	+	Living in harmony with nature
	−	Ineffectual endeavours to live in harmony with nature
WLE*	+	Cultivation of more appropriate (environment-friendly) consumption patterns / "Caring for the Earth"
	−	Fadism / Cultism
WLR*	+	Creation of opportunities to fulfill aspirations and identity needs / Affirmative action / Appropriate lifestyles
	−	Alienation / Substance abuse
WLT*	+	Cultural development
	−	Cultural homogenization / Cultural imperialism
WTE*	+	Low-resource living and production patterns / Sustainable agricultural development
	−	Subsistence agriculture / Grinding poverty
WTR*	+	Creation of new economic opportunities / Improving conditions of work (quality of working environment) / Allocation of resources for development
	−	Ineffectual development initiatives / Structural adjustment "without a human face" / Informal economies

18

Figure 4

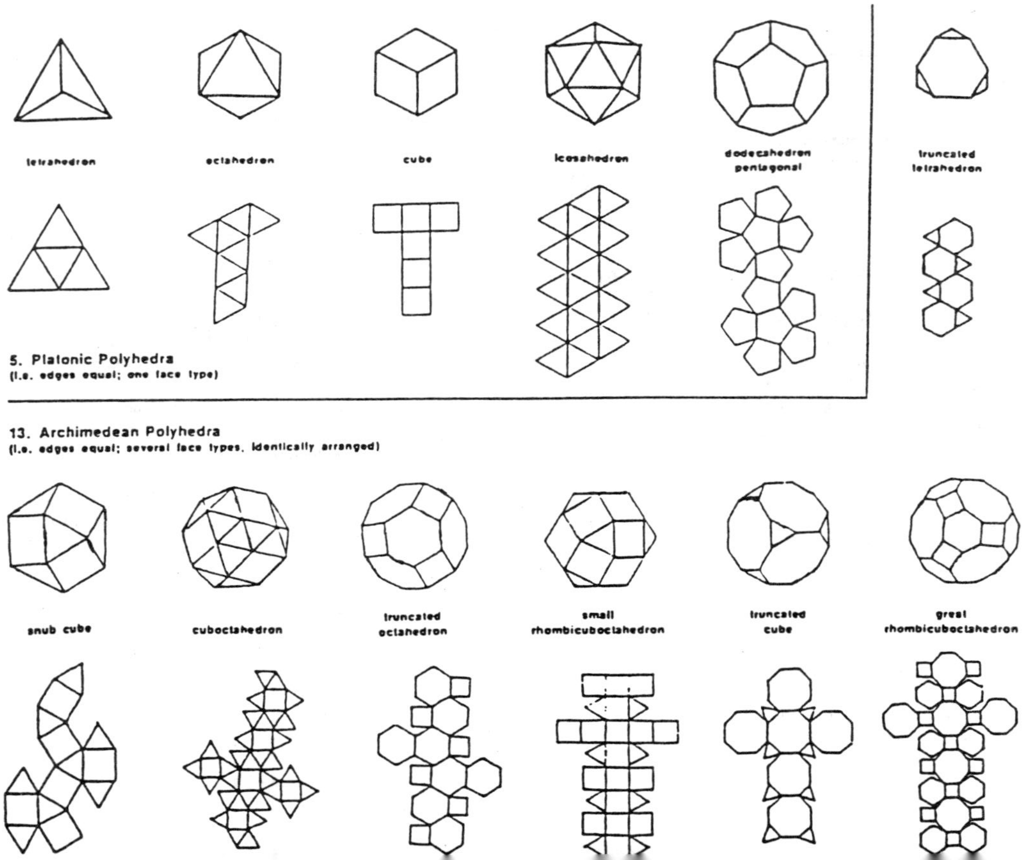

truncated
tetrahedron

dodecahedron
pentagonal

icosahedron

cube

octahedron

tetrahedron

5. Platonic Polyhedra
(i.e. edges equal; one face type)

13. Archimedean Polyhedra
(i.e. edges equal; several face types, identically arranged)

great
rhombicubeoctahedron

truncated
cube

small
rhombicubeoctahedron

truncated
octahedron

cuboctahedron

snub cube

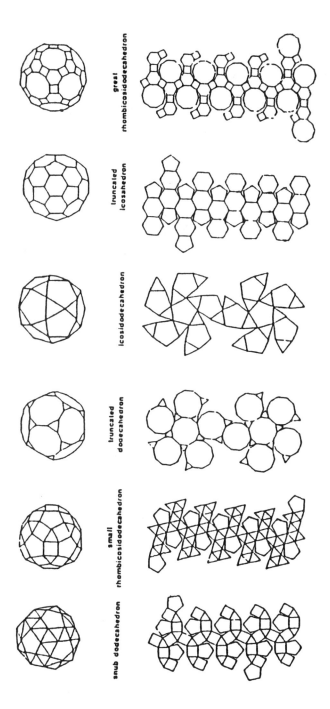

Patterns of Sustainability -- from 2D to 3D

Possible maps of globally sustainable bargains basic to the design of tensegrity organizations. The diagram in Figure 2 is based on the icosidodecahedron in the bottom row.

Figure 5

Spherical representation of icosadodecahedral network

of Figure 2

Illustrating the emergence of global qualities interrelating local areas

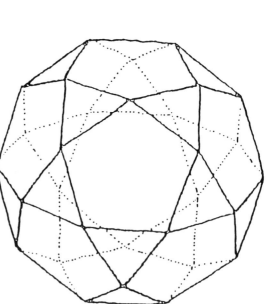

Figure 6

Vicious cycles and loops

A simple three-loop is:

Disaccord (#PF5532)

Uncertainty → Fear
(#PF6438) ↗ (#PF6030)

Four-loops:

Limited historical
method (#PD3774)

Historical
misrepresentation
(#PE4936)

Failure to learn
from patterns of
history (#PF1746)

Commemoration of
dishonourable historical
events (#PF4422)

Unethical practices
of government (#PC0814)

Official
secrecy
(#PC1812)

Criminal violation
of civil rights
(#PD8709)

Crimes committed
during civil unrest (#PE11179)

A five-loop:

Rumour (#PF5596)

Alarmism (#PF4384)

Fear (#PF6030)

Inadequate structures
for communication (#PF2350)

Excessive state control of
communications mass media (#PD4597)

A six-loop:

Living alone (#PF3089)

Mental disorders of
the aged (#PD0919)

Fear of death
(#PF0462)

Deluded
quest for immortality (#PF2142)

Unnecessary personal
consumption (#PF5931)

Excessive longevity (#PD5973)

A seven loop:

Social neglect (#PB0883) ← Ignorance (#PB5568) ← Educational wastage (#PC1716)

Vulnerable children (#PD0513)

Drug abuse by adolescents (#PD5987)

Marginalized young people (#PC1946)

Social disaffection of the young (#PD1544)

Two six-loops linked by a common problem into a figure-of-eight:

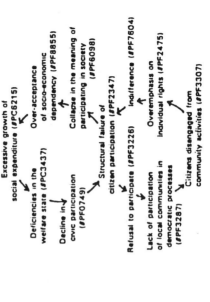

Excessive growth of social expenditure (#PC6215)

Over-acceptance of socio-economic dependency (#PF8855)

Collapse in the meaning of participating in society (#PF6098)

Deficiencies in the welfare state (#PC3437)

Decline in civic participation (#PF0749)

Structural failure of citizen participation (#PF2347)

Indifference (#PF7604)

Refusal to participate (#PF3226)

Lack of participation of local communities in democratic processes (#PF3287)

Overemphasis on individual rights (#PF2475)

Citizens disengaged from community activities (#PF3307)

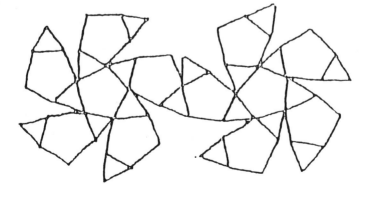

Hellmut Löckenhoff
FH für Technik, Esslingen; Berufsakademie Stuttgart

Modeling Knowledge
for Sustainable Environment Balance

Es geht also um die Rolle von Analogien
in der wissenschaftlichen Weltbeschreibung. G. Grössing
Das Wort ist nicht Zeichen, sondern Bedeutungsknoten.
J. Lacan

1. Introduction

Environmental problems have rapidly grown. They appear to be fundamental for life, highly complex, inexorable and mostly irreversible. They have also obtained a nearly all-embracing extent and impact. Concerns extend from garbage disposal and water pollution to economy, climate and electromagnetic interference. Methods applied range from recycling, all kinds of economy measures, efficiency and effectiveness up to cybernetics and AI for control purposes. Ecology, as becomes obvious, cannot be understood, let alone controlled, if not holistically as a densely meshed network; hitherto discrete approaches to environmental policy overlap and grow together. Without a sufficient overview of the general context detailed action cannot be designed. It presents the paradigmatical case for a sophisticated knowledge order. Using a phrase coined by the distinguished philosopher and mathematician V.V.NALIMOV: Ecology symbolizes the primary relationship of 'man in his world' to his world and, not least, to himself. As the adjectives 'global' and 'historical' indicate, the environmental space covers actually the planetary system and virtually the cosmos. The span of time goes back to prehistoric epochs and follows the arrow of time far into the endangered future of the spaceship earth. In respect to their control, limiting factors and absolute barriers are given. It is impossible to confine them to a deterministic, let alone a mechanical domain. Thus there are no one-layered rational or hermeneutical answers. Basic environmental processes cannot be influenced otherwise than longtime, the middle range counting in decennia and centennia rather than in years. Pollution effects, climatic change, demographic phenomena or energy consumption may serve as examples. In contrary transitions, catastrophic effects may show within a very short time, counting in years down to days. Examples refer to the dying of forests, to epidemics or the explosion of insect populations. Catastrophic short term effects obviously may also occur in the political, social or economic area. Modeling our patterns to understand the world must closely and precisely incorporate these challenging qualities. Our ability to successfully design an appropriate KO decides on the *options* we open-up - or close - for our future development.

Hence, to cope successfully with such complexity Environmental Knowledge Organization (EKO) acquires a crucial role. The world to be observed and controlled has to recognized first, to be ordered and to be organized accordingly. The entire networks mentioned are objects of knowledge organization (KO), in particular EKO. The task calls - remember Ashby's law of requisite variety - ideally for an EKO system of the order n + 1. It requires, moreover, insight into the material contents as well as into the intentions, the chances and limits of environmental control. Any attempt has to reconsider the basic preconditions of KO and EKO, of modeling models of our environment, of our world. EKO must, not least, be agreeable throughout society, communicable and transferable into institutions, measures and processes within an evolving learning cycle.

2. On Basic Parameters of Environmental Knowledge Organization

To know is to link new information with existing knowledge and to gain *insight* beyond the mere re-cognition of facts. In case of environmental knowledge an organized information base has to be provided as to *control* the mutual interactions between man, his artefacts (the Third World of K. POPPER) and his natural environment. Resting on which informational order this is done and employing which methods, appears paramount for the sustainability of environmental policy on all levels. In addition, the time horizons must be both, *on sight and longterm*. Operation cannot be linear, but must be *systemic*, taking into account non-linear processes as well as possible *side effects* on other, possibly remote parts of the system. A sustainable policy by necessity quests for such a systemic approach in all phases of environmental control: analysis and prognosis, goal setting and decision, planning, realization and control of the results. Improving EKO as a base for the control in itself acquires the quality of a helix, that is of *environmental learning*.

Any *equilibirium* attained will be of floating nature, constituting a temporary balance only for a given state and on an actual level of the *evolutional/learning process*. Sustainability itself is subject to change, occurring incrementally or in qualitative leaps. Insofar sustainability means continuity not so much in operations but in *guided change* within a strategy. Qualitative changes may not interrupt but modify the longterm development. They thus should preserve the basic potentials (as in *metamorphosis*) and open new fields for further evolution.

General considerations like these may seem trivial. They are outlined here in brief to accentuate the fundamental requirements the design of EKO has to meet. This rough draft may also draw attention to the multitude of aspects included - e.g. of structural, dynamic and systemic levels - from actual operation to metatheoretical aspects. EKO has to reconsider the basic models, which govern our perception, our understanding, evaluation and intentions towards our environment. Not least, concommittantly, we build the models of ourselves, of our relations to our world and of our personality. In short: we critically have to quest our human 'reality' perceived and acted upon in its widest extent. On stake are also the underlying systems of evaluation. They in the first line determine what we see and what we do not become aware of. More important, they also decide to which phenomena, facts, goals and ethical norms we assign which kind and degree of importance to.

EKO is the result of our picture of the world and, in reverse, constitutes the ordering structure, the organization to control that world. Hence any focus on EKO has to begin with the *conditions of modeling* appropriate to sustainable environmental balance. Every aspect of human co-existence and co-evolution will be involved.

3. Tasks to be accomplished, preconditions to be observed

This paper will restrict itself to three seemingly simple questions and try - if only provisionally and partially - to attain aspects where answers may by attempted. Approaches will combine issue based systematic order with formal listing for better pragmatic transparence.

3.1 As any KO, EKO, too, is implicitly a quid pro quo, is a means to an end. Thus, to what *intention* is EKO directed to? For what *control purposes* must EKO provide appropriately structured knowledge? What does appropriateness include in terms of *controllability*, e.g. towards depth analysis of causes, interrelationships, possible longterm developments and, in particular, sustainable if floating equilibria? Where do the fundaments of, the tasks behind and the feasible modes of EKO lie?

Analogous to the philosophy of industrial controlling environmental control includes all phases: analysis and prognosis, goal setting and decision up to planning, realizing, evaluating intermediate results, comparison with planning objectives and, finally, decision where and how to intervene. The first phase, prognosis, undertakes to choose the possible fields of actions and to define the *optional strategies*. Though alternatives are, by all practical means, numerous and rather costly even when provisionally to evaluate, *simulation* necessarily must be employed. This holds true in particular when e.g. technology/strategy *assessment* proves advisable in case of environmental consequences. Interpretation asks for *priority scalas*, for *indicators* signalling trends to expected e.g. in *early warning* information systems. On which criteria, short or long term, e.g. the success or failure of the "Green Dot" is to evaluate? Rather similar requests are put by goal setting and goal decision, by designing planning alternatives - simulation here proves indispensable - and in order to evaluate the steady flow of informations on operational/strategical performance. How to place the interfaces where intervention optimally should take place, which intervention rhythms and spans of time given? In comparison to this *procedural* the *material/factual* side only superficially appears more easy to handle. Consider for example only the ordering of garbage in respect to disposal/recycling or the constructional order of a more complex product in respect to its recycling after use - e.g. the automotive industry. Applied EKO has to find solutions.

3.2 Models provide the framework, wherein EKO will constitute the control order system. Are there, and if so, which *basic rules of modeling*, concerning the aspects of form as well as those of content and their interferences? Referring to the much quoted change of paradigms, scientific and profane, it does not suffice to analyse and compare the changes taking place in the prevailing models in physics and biology, in cognition, experience and evaluation. The thoroughness of change quests the *fundaments of ordering/structuring*. It affects the physical level, the fields of life and the realms of consciousness alike. Is there a universal base, a common structure for modeling on all these levels? Does such a shared pattern contain by necessity an environmental factor, is any KO insofar an EKO?

Measuring, that is, what to measure, where, by which criteria and which way, is only one of the more plain problems arising. On what *indicators*, in what aspects and with what explanatory power may the assessment of the ecological situation in general rely? Or energy consumption per head, or relative space used for transportation? And how could a price be defined which would express - mark the fuzzy term - the ecological cost or the 'ecological thruth'? Can environmental, social, ethical aspects be coupled for a comprising evaluation, and if, under which conditions? How do hierarchical and *networked* paradigms go together, how can horizontal and cross-thinking be combined? These and related assignments have to be transferred into modes of modeling.

3.3 These quests lead to *metatheoretical* research on KO. Though not always obviously and/or instantly, the insight into the rules of modeling and knowledge ordering will not only facilitate the badly needed transparence of existing *theories on modeling*. Metatheoretical knowledge will in turn influence theory building, modeling and classification. It will help to detect inconsistencies, wrong elements a.s.o. in still prevailing EKO and will hopefully hrlp straightening them out. First steps as e.g. in the cognitive sciences can already be observed and serve as examples. Moreover, growing insight into metatheoretical processes will promote sophisticated modeling and EKO in actual cases and gain *hermeneutical* qualities.

Hermeneutical modeling will facilitate the steady *adaption* of KO to changing pragmatical needs. It will, hopefully, encourage us to connect and integrate already existant modeling approaches into an *integrating network*. Models as the relating KO from different sources have to be *coupled*, e.g. when following an overall 'least cost' planning. It will be the base of the urgently required - because still non-sufficient - *interdisciplinary coordination* of differently defined ecology sciences. It is necessary to state more precisely crucial *concepts* of *sensitivity, sustainability, resilience* a.s.o. Still more intricate turns out to be the incorporation of apparently exotic subjects into the EKO and ecological models: so the *handling of visions* and similar creative instruments as *archetypes of human thought*, and, of course, *epistemological* topics and as the use of *analogies, of association and bisociation*. For orientation a reference to the most general and basic *concepts* often seems necessary: to *symmetry and asymmetry*, to order, to similarity and to differentiation. This trend can be found, latent or open, within and behind approaches like that of selforganization, of synergy and of related typically both formal and procedural concepts. It is not mere coincidence when we experience a *third wave of the mathematization* of the world and the algorithmization of KO. Approaches concern, analogous to the above mentioned more general trend, such basic concepts as number theory, in particular prime numbers, and the basic relations SPENCER-BROWN dealt with in The Laws of Form. In addition, the domain in toto is affected by the omnipresent challenge how to *cope with complexity* in a non-reductive, creative mode.

Questions and responses to 3.1 - 3.3 overlap and presuppose each other. They sketch a general framework for questioning and problem solving rather than a structuring principle for the following paragraphs. The result stresses the necessity of a thorough reconsideration of basic assumptions behind and of principles applied to modeling and EKO. - How, in particular, are the various aspects of man's relationship with his world affected?

4. The Challenge to Science and Knowledge Order by Ecologic Sciences

Although not in every aspect (a fact easily overlooked), science nevertheless *constitutes the main generally accepted KO* and thus also the EKO of coping with tasks. Science undertakes to analyze and solve technical problems as well as those of society, the individual and, of course, of ecology. Judged against the historical development the ecological problems we are facing appear as the aftermath of the Western way to comprehend the world. From roots sprouting 2000 years ago and even earlier sources, it emerged into its now virulent shape, initially in the Renaissance. Unique in history, it organized knowledge predominantly along parameters of *mechanistic/rational control*. In consequence, the linearity of this in common practice still valid model caused serious disturbances in the social and the ecological equilibria. Either ironically or only consequentially, this very scientific approach is employed to meet also ecological challenges. Very early, however, the insight was gained that expelling the devil with Beelzebub was not appropriate to such a most complex phenomenon. In particular, indigenous domains of non-determinable character such as those of chance, evolution, of voluntary elements within human interaction with environment and ethical questions appear. If reluctantly, it is accepted that - if only to prevent further damage - the accepted scientific models and the rules of rational modeling need be complemented by nonlinear, 'fuzzy' models.

The attempt to devise more sophisticated and more open EKO models contributes to the general endeavour to *restructure with science* also KO. Not that there are no obstacles: methodological inconsistencies as well as the defense of hereditary fiefs of KO. The quest for change is but naturally felt as a threat against *continuity and stability*. Even if there is consensus, there is often no cooperative action. Scientific deficits become obvious, when ecological research poses with new questions new quests. These may refer to not yet existent factual aspects within disciplines and to methodological difficulties for interdisciplinary coupling KO, to methods of research and the evaluation of results. Latent insufficiencies of existing *paradigms* become virulent; new ones may be still fuzzy and not yet accepted. Science gives the impression of being far more in a state of transition that it is normally accepted by the scientific community. The prevailing rather critical discourse concerning the tasks and the efficiency of environment research institutes provides but an example. With the critical situation of man in his world the requests of man to the scientific community as an instrument to overcome problems and to gain welfare have dramatically changed towards *sustainable effectiveness*. As will be shown later, science must contribute to a new model, a new KO which enables mankind to cope sustainably with impending transitions, rapid changes and catastrophes. There is no time left, so science had better be extremely expressively pragmatical.

5. Principles and Modes to Construct Knowledge Systems
to Structure the Environment

To remind the topic of this paper and the task: knowledge order must be modeled in such a way, that it helps to constitute and to elaborate, to design and continuously to adapt to the actual and to future needs. It has to account for actual *protection* against catastrophes as well as to preserve *potentials* for future options and general evolution. As already pointed out, 'Environmental Balance' serves but as a paradigmatic label for a thorough, comprehensive and worldwide consignment.

In terms of modeling KO and EKO a threefold effort is asked. First, KO research will outline the constraints of modeling as a formal, physical, physiological and cognitive/ experiental process. Paying respect to the actual facon de parler using the lingua franca of today that may be called 'back to basics'. Second, modeling abstracts from the de facto unlimited variety of possible aspects of reality. Modeling and/or its constitutive KO simplify the particular 'reality' we observe and react upon. Uno acto with classifying, as a radical contructivist may say, we construct our reality. Or, as the systemic approach would state, potential variety is reduced concomitantly when priorities are set, embodying intentions and restrictions. It predestines also the ways favoured when interacting with the environment. It will be necessary, then, to inquire the methods of modeling as to the ways they restrict. Mere mechanical reduction is opposed by purposeful setting of limits and goals, providing for new potentials for future development. - Goals, third, emerge from values, from ethical attitudes; they express *meaning*. In consequence the impact of ethics, from untilitaristic consideration to ideologies, must be questioned.

Despite all efforts to gain a more comprehensive picture of man and his world the prevailing models of both are still essentially performed by ideas originating from the early natural sciences. Though more sophisticated and not quite as onesided as before, they still set narrow borderlines for open, creative thinking. Worse, they enhance a policy which not canalizes but immobilizes human intuition, intentions and chances to help oneself. Economy and social sciences, respectively economic and social policy deliver striking examples. Cherishing a rather specialized picture of man as subject to control, the scientific disciplines contribute to this very effect. From an approach highly specialized and thus restricting KO and EKO to partial aspects, they originate working models, which, as homo economicus or homo sociologicus, have gained ill fame. The obvious countermeasure appears to be interdisciplinary cooperation. That would, however, presuppose models and KO's and EKO's which are compatible. Which further quests, since models have to be specific to be effective, that they must be derived from a shared, necessarily rather formal model following shared rules of differentiation. Which, in turn, asks to reconcile basic assumptions, basic KO's ordering man's indigenous structures within his environment. Fortunately the scientific disciplines are, by massive impact from reality, rejuvenating from inside and open to redefine the interfaces that separate them or, hopefully, connect them which each other and the vision of science as an integrated body of ordered knowledge. Here the argument turns circular: every discipline must contribute to the basics of a picture of man in his world, from mathematics to psychology. This issue, however, will be accomplished only within a shared framework of those disciplines. The challenge to KO/EKO may be seen as by discourse to change this barren circle into an evolutional helix.

Which modes and rules should modeling follow, if it wants to support evolutional qualities? As might be learned from the preceding discussion, modeling must be open, flexible, comprehensive; it must be designed as the backbone of the evolutionary process itself. To our present state of knowledge these requests are met only by the well known variances (!) of the systems / the systemic approach and, concerning procedures, by the hard and in particular by the soft systems method (SSM).

Who dares to design a model, a KO of how to behave rightly, in particular a EKO, as a base for *norms* governing the interaction between man and his inner and outer environment? In fact, those approaches are encouragingly already emerging. On the material level they may be detected in the protective legislation to save as much 'natural' environment as possible as to prevent the deterioration of the life sphere. Applied strategical ethics as those encircle the fields, where voluntary *ethical restrictions* or aims may be set. As for procedural rules of systemic goal setting, task performance and problem solving, essentials can be found in the recommendations of the UNO. They consign precautions before repair, soft methods before hard ones. On the next level natural philosophy, philosophy of natural sciences or philosophy of nature, explicitly or within the (hiven-) anthropologies (such as bio-sociology or bio-psychology) is contributing to the discourse towards a more sustainably fitting picture. Finally, in the domain of beliefs, religious, secularized ideologies, or just wellmeaning movements from green to red-green, in the long run nurture the hope for at least more clearcut pragmatical attitudes. It is but paramount to hold the discourse wide open.

In neither case, concerning interdisciplinarity or ethics, a thoroughgoing structure nor uniformity of expression can be intended. Contrarywise, inconsistencies, gaps and contradictions constitute a necessary cause for continuous learning discourses. Since elaborated elsewhere, it is sufficient to recall that any discourse, any evolutional helix possesses the qualities of a *learning process*.

6. Concepts of Ecological Policy and Eco-Controlling

Ecological challenges have to be dealt with when employing appropriate KO/EKO, incorporated in models. Modeling follows evolutionary principles of open learning. The actual scenery, however, where ecological policy has to be made and eco-controlling to be practiced, is formed by the present state of our world and the actual local situation.

Classifying and summarizing the eminent features it may be permissible to distinguish, if grossly simplifying, three clusters of phenomena and principles. The first one to identify the present inconsistencies and their results is called the *A-and-I cluster*. It describes the transient loss of established orders - values to attitudes and institutions- and its results we suffer from. Within a different context and using different terms the phenomena, the causes behind and the possible impact recently have been inquired by P. ATTESLANDER, notably 'anomy', and by a study on 'Values and Public Policy'. The latter tries to cover the whole societal-political development from roughly 1945 to this moment in the USA, allowing analogous application to the European and German situation. It was edited as the result of research ordered by the US government by H.J. AARON, T.E. MANN and T.TAYLOR; the most interesting part being that on the Dynamics of Economy and Values, Welfare and Society. Both publications shed light on the preconditions of our cluster, which connects *anomy* that is the lack of shared and accepted values, and, in consequence, the loss of accepted norms of social behaviour. *Aporias* fail e.g. to set strategies goals, accomplish longterm issues and solve problems especially in the demographic, societal, geological and environmental domains. The often quoted *inadequacy of institutions* represents but an example for the insufficient adaption of nearly any societal sector to meet the challenge of transience and future. A 'D' for 'des-' in, for example, des-orientation, might be added.

The indicating, contra-indicatory principles to construct models and procedures may be collected into a 'CO' cluster. Co-g-nition, that is knowing the context around and behind, contributes the basis for co-operation, co-action and co-evolution. The meaning of 'CO' is not confined to the individual, to the community, the nation or worldwide coexistence. 'CO' explicitly and foremost means shared existence with the world as the sphere of life in its widest understanding. While the A-cluster predominantly signifies a loss, the CO - cluster, as an evolutional answer, carries principles of factual and ethical command as expressed for instance in the Sermon on the Mountain. The factual, strategical and voluntary request contained will be the scala to evaluate and to judge any political measure communal, European or worldwide. Moreover, they set the nodes and the limits for sustainable co-acting.

The *'SL' cluster*, finally, marks the formal framework for structuring organization and procedure: *Systems, and Evolutional Learning*. The learning, evolutionary KO/EKO has been dealt with elsewhere. Its underlying principle holds valid for any model, any mode of goal setting, decision and controlling, of any conscious design towards agreeable, consensual behaviour. The SL cluster will include what may be called the 'meta' and 'para' approach. Close to what was termed cross-thinking - 'Querdenken' in German - it signifies research transgressing the established and agreed limits of well-founded hypotheses into the realm of speculations, however critical and responsible they may be.

Western culture, if one-sided, has been so very successful because it confined itself to rigorous rationalism. To meet the obvious and the latent challenges we certainly must foot on the results, the methods and the models of conventional science. They alone, however, prove, neither sufficiently powerful nor effective in the sense of the CO and the SL cluster. The challenge originates in essence from the reality 'beyond' the reality we shaped with traditional science, and thus effectively excluded any other reality, just ignoring it or waving it aside into the sphere of religion, ideologies or crackpots. It becomes urgent now to be open for unconventional approaches, models and procedures even in KO and epistemology. This should be ventured and approached cautiously, responsibly and very deliberately, but it should be attempted also consistently. Consistently not in the last place, since in doing so we return to the spirit of the founders of our now conventional science.

7. Modeling Environmental KO Towards Evolutional Teaching/Learning

Ecological policy, ecological controlling inevitably takes place in the societal dimension. An even authoritarian attempt will have to allow for the mediation of information; knowledge and insight, of underlying philosophy, principles agreed with, models, methods and procedures to be followed and bench marks to be observed. The CO as well as the SL principle ask for the steady transfer of order impulses and order control through open communication to evolutional learning. To achieve optimal results *involvement and commitment* are desired - *intellectual, cognitive and emphatically also emotional.* Both the energy and the innovative ideas preponderantly originate bottom-up, not top-down. Without a clearly formulated framework and strategies, but also without a continued open discourse any policy is doomed to be ineffective. It will lead to dead-ends where options for the future should be developed.

More on the pragmatical level, ecological policy must be *communicable*. The aspects are manifold. Communication presupposes confidence. Confidence relies on transparence and trustworthiness of objectives and the reason supposed to be behind, presupposing a similar picture of the situation and its challenges within the partner's head. The emotional understanding, the base for involvement and voluntary commitment, is retrieved or missed here. Emotional involvement furthers insight. A prerequisite for such active acceptance is the factual, material understanding of what is communicated and discussed. To sum it up: how can the design of the EKO further understanding and communication? What principles should the EKO pursue, or: what basic rules of *building a joint framework of understanding* by steady communication should be followed?

This question so far has found no fully satisfactory answer. Partly, because after a good decennium of intensive ecological/environmental research, the ability to procure information and to get research results is not equalled by far by the ability or intention to apply them pragmatically to actual tasks. Not only therefore *issue oriented information* poses problems. The landscape of knowledge itself changes continually with increasing information. Similar difficulties arise when *consultative information* is to be given to serve actual purposes or to accompany long-term projects. A most important target for environment information is prescribed, for example, by *technology assessment*. Government is asking for geographical and geological background information, known as the request for *policy assistance*. Again the example: how should household garbage be classified? In order to be disposed of, recycled, prevented, ruled by laws, as a communal task, as an individual responsibility, etc. etc.?

The problem centers on the task to *secure a shared fundamental* understanding - necessity, philosophy, factual networks, modes and instruments, impacts and consequences a.s.o. - where actual understanding can be built upon and which can, following the development, be processed and kept at the current state of knowledge. In other words: a process of continuous environment and ecology learning has to be established. Learning begins if not in the *family* then at the *primary school*. H.v.HENTIG has only recently elaborated the school as a polis, a learning and exercising ground on how to be a good citizen. An enlarged and deeper understanding is required to accept and act according to *policy measures*. Methods - see J. ZELGER - have been developed to find out what people are thinking e.g. about a societal phenomenon and how they *evaluate politically* enforced measures. Using these instruments, governments may find out the best way to design regulations and to communicate them to citizens with an optimal chance to gain acceptance.Putting aside communicative problems as that of political/social correctness and other ideological distortions: what can the science and the daily pragmatics of KO/EKO contribute to ecological communication and environment balance? How can social resonance and commitment, can individual emphasis actively be supported? Oviously by opening even wider for interdisciplinary cooperation with centers of environment research and with political instances. Further, concerning the basic communicative aspects discussed, in establishing work groups. Finally, but not conclusivley, actively transgressing the own professional areas for joint ventures.

Note. By deliberate intention the paper transgresses in some of its arguments the base of accepted science. It does so not without considering the plausibility and usefulness of a further discourse. Restrained to main features woodcut qualities could not altogether be avoided. These may be properly differentiated in the actual case.

Selected References

(1) Aaron, H.J. et al. (Ed.): Values and Public Policy. Washington: Brookings Institution. Report for the US Government 1994.
(2) Atteslander, P. (Hrsg.): Kulturelle Eigenentwicklung. Frankfurt/M. New York: Campus 1993.
(3) Grössing, G.: Das Unbewusste in der Physik. Wien: Thuria & Kant 1993.
(4) Hentig, H.v.: Die Schule neu denken. Stuttgart: Hanser 1993.
(5) Löckenhoff, Hellmut: Systems Modeling for Classification: the Quest for Selforganization. Knowl.Org. 21(1994)1 p. 12ff
(6) Nalimov, V.V.: Time, Space and Life. The Probabilistic Pathways of Evolution. Philadelphia PA: ISI Press 1985.
(7) Zelger, J.: GABEK VII Zur qualitativen Auswertung sprachlicher Äusserungen. Preprint 27. Projektberichte, Universität Innsbruck: Institut f. Philosophie 1994.

Elmar A. Stuhler, Marjan Vezak
Technische Universität München

Basic Concepts in the Organization of Environmental Problem Solving (Cognitive Information Processing)

0. Introduction

This paper breaks down into three sections. the first deals with the principal concepts of environmental policy, the second with those relating to the organization of cognitive information processing in order to solve dynamic, complex environmental problems, and the third with what would appear to be the principal concepts for simulation models which can be used in order to simplify cognitive information processing and render it more efficient.

The chief aim of this procedure, which utilizes a number of disciplines, is to establish a link between the natural sciences, systemic cognitive anthropology and systems theory and their application in simulation models. The authors are of the opinion that it is demanding too much of ecology to seek to develop it into a new, interdisciplinary science. The primary function of ecology is to describe nature.

1. Environmental Concepts

1.1 Sustainable development and the need for precaution

Since the United Nations Conference on the Environment and Development was held in Rio de Janeiro in June 1992 the leading concept for the environmental policy of the future has been SUSTAINABLE DEVELOPMENT. This concept includes the NEED FOR PRECAUTION in order to anticipate and prevent undesirable developments.

The concept of sustainable development requires that economic, social and ecological development be regarded as a unity. This is an essential condition for the responsible management of natural resources since it precludes any of the three aspects being played off against the other two. If this does happen there is no assurance that human development will continue in the longer term. Only if civilization in its various forms is thoroughly integrated into the natural network that supports it can progress be hoped for in the longer run.

1.2 Carrying capacity

The concept of CARRYING CAPACITY is closely linked to that of sustainable development. It signifies the strength of ecological systems which economic and social processes must take into account. The carrying capacity of natural systems necessitates a change in what is generally thought of as economic, technical and organizational PROGRESS. From now on progress can only be used to denominate that which the natural environment is able to support. This means that modern concepts of progress will have to be rethought.

1.3 Recycling

If industry wishes to contribute to sustainable development, carrying capacity and the need for precaution, it will have to lay great stress on RECYCLING. Manufacturing processes are closely bound up with and influence the various natural cycles. A cyclical economy seeks to achieve economic and social development while preserving ecological systems.

1.4 Interlinking, flowing equilibria
What has been said so far indicates that the ecological problems that confront us today are problems of INTERLINKING. Ecological, social and technical systems need to be appropriately linked with one another and the relevant ecological systems if ecological balances are to be preserved.

The FLUID BALANCES, which within certain limits constantly re-establish themselves in nature, help to compensate for changes brought about by civilization. However, developments brought about by man need to allow for their temporal rhythms. Here it is vital to ensure that the carrying capacity of the environment is not exceeded, as otherwise no sustainable development will be possible and effects will be irreversible.

1.5 Ethics, anthropocentrism, essential meaning of nature
The GLOBAL INTERLINKING of the problems we have outlined cause ETHICS to take on particular importance, as regards both the acts of individual persons and government policy concerning the environment. It would seem immediately obvious that traditional anthropocentric ethics are not particularly appropriate in this context of global interlinking. We need to develop a holistic ethos that because it takes account of the whole accords with the complexity of the human environment and extends beyond the ANTHROPOCENTRIC APPROACH, allowing for man's responsibility for the natural environment, which has its own inherent significance. The INHERENT SIGNIFICANCE OF NATURE in the sense of non-human nature calls for specific moral standards of its own.

1.6 Ethical judgements, areas of conflict
Embedding man-generated systems and developments in the natural environment and adjusting them to the flowing equilibria to ensure continuing survival leads to the emergence of three areas of ETHICAL JUDGEMENT: As technical progress is achieved man becomes increasingly responsible for the natural environment. Moreover, there is a need for a just coexistence between human beings. Finally there is man's responsibility for himself. What kind of basic principles and criteria contribute to sustained development despite the THREE AREAS OF CONFLICT? What is involved is the encouragement of an overall sense of responsibility, the overcoming of social polarities and arriving at a consensus. This requires a clear presentation of factual situations and realistic assessments of risk without over- or understatement. Ethics should be governed by principles whilst also being susceptible of practical application. But what is ethically defensible in individual cases can only be decided on a case by case basis.

1.7 Manageable priorities and basic rules for action, environmental ethics
It is possible to take a number of cases and from them gradually evolve MANAGEABLE PRIORITIES AND BASIC RULES FOR ACTION. This is the task of an application-oriented system of ENVIRONMENTAL ETHICS. Such as system must concern itself particularly with methods for weighing up ethical pros and cons. Man has to analyse and decide, bearing in mind the need for precaution, which is the smallest of several evils. This is the basis for a constantly evolving optimization process and progression towards the best possible solution.

1.8 Protecting the environment, safety distances and free spaces
PROTECTION OF THE ENVIRONMENT is one of the responsibilities of government and must stand as a stated aim for all government authorities. It must take account of the need for precaution and not confine itself to post factum remedial measures.

Sustainable development and the various requirements that go with it must be a task prescribed by government, whose particular responsibility it is to provide for safety distances and free spaces to prevent risks to man and nature which it will probably not be possible to assess until some time in the future.

1.9 Setting environmental targets, environmental indicators
SETTING ENVIRONMENTAL TARGETS includes formulating action-oriented specifications for desirable environmental quality in material, spatial and temporal terms. One could also call this setting target limits. What is important is that while targets are being evolved their interdependence is also becoming clear. The evolving of

partial goals is essential for the application of the sustainable development and carrying capacity concepts.
The ENVIRONMENT INDICATORS used in translating these various into practice include the CRITICAL LOAD CONCEPT and what are known as CRITICAL STRUCTURAL CHANGES and CRITICAL LEVELS. This last can be used for REDUCTION STRATEGIES for reducing the degree of environmental pollution - which may differ according to the needs of each region - and achieve QUALITY TARGETS such as reducing eutrophication.

2. Concepts for the Organization of Cognitive Information Processing when Dealing with Complex Environmental Problems

2.1 A few introductory considerations
Let us assume that John Smith is the person responsible for matters affecting the environment in a certain district including a city with 250,000 inhabitants. John Smith is doing his best to achieve the environmental aims for his district. However, he is faced with the problem that so far no one has been able to explain to him exactly what is to be regarded as sustainable development, carrying capacity etc. in individual cases. What does it mean, for example, when a factory belonging to a large chemical concern applies for permission to expand its manufacturing plant? John Smith's superior does not discuss applications from industrial firms with his staff, although this would be absolutely necessary for their efficient processing. There are around 19 laws and regulations governing matters connected with noxious emissions and they are far from easy to interpret. John Smith regrets the absence of discussion at the level of his own authority as to how central government directives for dealing with such applications are to be handled in practice. Those left to do the job without guidance often find the demands being made on them excessive.

Having briefly outlined the problems of work organization in one area of a local authority we must ask ourselves how John Smith can proceed to organize his cognitive information processing in his field in order to optimize his decisions. To do this requires that we consider the human mind as a computational system.

2.2 Complex problem-solving
Much of human activity can be viewed as the solving of complex, dynamic and more or less UNDEFINED PROBLEMS. More recently environmental problems have assumed major importance. Problems or situations are described as undefined when there is a lack of clarity as to the goals and the means to be used to attain them. UNCERTAIN PROBLEMS exist when developments or a sequence of events cannot be foreseen, as in the case of natural catastrophes, which usually occur suddenly and unexpectedly. Problems or systems are referred to as complex when they are the product of a plurality of factors. Many COMPLEX PROBLEMS are INTERLINKED PROBLEMS. They cannot be isolated from other complex problems, which carries the risk that solving one problem may aggravate another problem or even create new ones.

We talk of a problem existing when a person attempts to remedy an unsatisfactory situation and create a better one, which we call the goal situation. To do this one has to employ suitable means and overcome obstacles.

(a) If John Smith wishes mentally to work out the solution to an ecological problem he has to do so in steps. Research in this area is partly concerned with observing the actions of a PROBLEM-SOLVER - the visible part of the process. This becomes more transparent if the problem-solver discusses his mental processes aloud. In the case of ill-defined, dynamic and complex problems no single step is usually the key to the solution of the problem. The process involves finding a number of steps which are all "correct".
The difficulty lies in deciding which steps are needed to get from the initial situation to the goal situation. So that responsibility for the solution "is spread over the whole solution process rather than falling on the discovery of one or two key steps" (van Lehn 1989, page 527). This makes it important how John Smith organizes the cognitive information process when attacking his environmental problems.

We may assume that the environmental problems John Smith has to solve are not KNOWLEDGE-LEAN but KNOWLEDGE-RICH DOMAINS (cf. Newell and Simons landmark theory). Other knowledge-rich domains are, for example, the formulation of public policy or medical diagnosis.

It is not difficult to grasp the difference between someone like John Smith, who is an expert with years of experience behind him, and a novice. LEVEL OF EXPERTISE, therefore, is another important factor in the problem-solving process. When an expert has been confronted with the same or similar problems many times over, we may assume that he or she will quickly grasp the essence of the problem and recognize its pattern as familiar, and that he will draw the elements useful for solving the problem from his MEMORY - EPISTEMIC STRUCTURE in order to devise a solution.

Although John Smith is far from being a novice, the kind of undefined, dynamic and complex problems he has to cope with do not support the hypothesis that expertise permits recognition to replace the search for a solution. The type of cases he has to deal with oblige him to search for a solution. His expert behaviour is solving environmental problems is mainly determined by the HEURISTIC STRUCTURES of the mind. This means that the mind is able to find solutions through a LEARNING PROCESS which may be based on TRIAL AND ERROR.
Although knowledge in this area is still far from complete, we conclude from our own observations and those of others that the problem-solving process and the learning mechanism might well be as follows and that the main patterns of John Smith's intellectual process could be as described in the following section.

(b) We assume that if John Smith has to analyze a case he will at a fairly early stage begin to formulate ideas and expectation - what we might term a dynamic model of the problem-solving and learning process. Let us consider the most important elements of such a model.

(c) John Smith's model may be regarded as the basis for ACTION REGULATION. As we have already mentioned, it will be governed by his knowledge of the problem - his epistemic structure. The greater the lack of relevant knowledge, the greater the need to work heuristically. How successful John Smith is in this will depend on the heuristic structure of his mind. How well can he perform the environment search processes or use analogy where his information on a given case is incomplete?

(d) Vital for an analysis of the initial situation or the goal set John Smith has to deal with is a thorough ANALYSIS OF THE COMPONENTS AND DEPENDENCIES of a critical variable. We consider a variable to be critical when there is a gap between the goal and the initial situation. An analysis of components and dependencies is necessary if it is unclear what the components of a critical variable are, what the structure of the system is and what are the interactions between the various components. Take the example of the large chemical concern in John Smith's area that wishes to expand its production facilities. The case is a sensitive one in that the company has been operating in the area since the last century. The components of the case need to be carefully analyzed to clarify all the various aspects. It is also necessary to analyze and re-assess the dependencies known so far and to evaluate new ones. These dependencies include the expansion of the chemical plant, the increase in pollution that this will bring about and the number of jobs that will be created.

(e) The analyses of components and dependencies are inter-related because both help to clarify the structure of the environmental problem (system). Part of the process of organizing cognitive information processing or of action regulation lies in determining the DEGREE OF DISSOLUTION of the components and dependencies. The more details that are available, the clearer the structure becomes - what Schroder et al. refer to as Differenziertheit or Diskriminiertheit versus Integriertheit. However, once a certain degree of dissolution is reached it may be useless to probe deeper into the various sub-units of the components. The same applies to the dissolution of the dependencies.

Memory plays an important part in this kind of analysis. The more pictures of components and dependencies it has stored, the easier the analysis becomes. Where there are no such memory pictures, another set of tools could be of help (DOERNER, 1983).

(f) Subordination, superordination and conclusions from analogies.
The process of SUBORDINATION is a search for sub-terms, which are more familiar to the problem-solver than the general term. Thus it answers the question of what type of pollutants are involved in pollution by the chemical plant.
SUPERORDINATION searches for a general term to include several sub-terms. It may be that one searches for a general term in an analysis when analyzing components and dependencies. Our chemical plant, for example, generates a number of specific pollutants for which John Smith has so far no general term. One might call then NEGATIVE EXTERNAL EFFECTS.

(g) Where the memory does not have the facts needed for an analysis of components and dependencies, the method of TRANSFER BY ANALOGY can be used. Let us define what this means. A complete transfer of knowledge rarely occurs because, for example, the knowledge of a doctor of medicine differs from that of an engineer. But knowledge can overlap (epistemic and/or heuristic structure). Even if the knowledge required to cope with two sets of problems does not apparently overlap, "There still may be a general transfer because problem solving in both domains may require an organized, methodical type of thinking. So training in that type of thinking in one task domain may give a subtle advantage in another task domain. For instance, learning to program a computer is often thought to increase one's ability for logical and quantitative problem-solving of all types (see, for example, Papert 1990)." (van Lehn 1989, page 556).

(h) PROBLEM SEARCHING provides an alternative method where the actor does not wish or is unable to make an analysis of sub-objectives by dividing a global goal into increasingly refined sub-goals because of the risk of losing the overview.
Both the formulation of sub-objectives and problem searching may be used to develop a hierarchy of objectives or sub-goals.

2.3. INTENTIONS and TIME PLANNING
involve the central question of how John Smith treats intentions related to his environmental problems. An intention results from the deviation of a critical variable from what the value should be. It involves the question of which matters should be dealt with first and how much time should be devoted to each. Thus it brings in the concepts of IMPORTANCE, URGENCY and PROBABILITY OF SUCCESS.

Importance depends on the degree to which a critical variable influences the objective. Urgency relates to the size, the speed and possibly also the acceleration of the deviation of the critical variable from the goal situation.

John Smith's probability of success or ACTUAL COMPETENCE in managing an intention is determined by his epistemic and heuristic competence as well as by the time available. Epistemic competence can also be termed "available operational knowledge"..
An expert is a person of high epistemic competence. Heuristic competence is a person's ability to carry through his intentions despite a lack of sufficient epistemic competence. This means that the method and means of solving the problem have to be worked out before they can be applied. How successfully this is handled will depend on a person's heuristic structures.

2.4 A final point
to be made is that we must not only analyse the side effects that certain measures will generate. We must also be aware that all cognitive processes are embedded in EMOTIONAL PROCESSES. Thus where a person does not possess sufficient actual competence there is a risk of fear and stress giving rise to panic reactions liable to influence the organization of cognitive information processing and producing a number of undesirable side effects. Control of emotions and a monitoring of one's own strategies

(organization) for information processing is very important and calls for a lengthy period of reflection. MANAGEMENT OF EFFECT is also important because "in order to continue to be effective ...time pressure combined with a restricted knowledge of the system may lead to worry and helplessness (Janis & Man,1977; White, Wearing & Hill, 1994). It may then become necessary to deal with these feelings first so as to remain able to act effectively. These difficulties are not met with only when tackling complex dynamic tasks, but they are likely to be particularly important in these cases because the role of control processes is more important (Doerner & Wearing, 1994, page 8).

It is clear that as a rule experts like John Smith will have better control over situations and internal processes than novices. Experts can work faster than novices, they are more accurate and therefore run into fewer difficulties. They also seem to be better at monitoring the progress of problem solving and allocate their efforts appropriately.

As we explained at the beginning of this paper, the purpose in each case is to identify the problem rather than provide the correct answer. Our brief overview of basic terminology needed when dealing with environmental problems is not complete. Our aim is to show the need for further efforts in this direction.

Everything we have said so far assumes that the problem-solver is able himself to "digest" the data he has to process. Where the flood of information and the complexity of the environmental problem or system is too great, it may become necessary to make use of simulation models. The third part of our paper contains some comments on this aspect.

3. Simulation Models as an Aid to Decision-Making

Having surveyed the various possibilities and limits of action regulation - or organization of cognitive information processing - we need to look at what means John Smith has available when the demands made on his own mental capacity for information processing become excessive.

Despite the insight than can be gained using the tools we have described, the human mind has only a limited ability to recall facts (variables) and usually encounters difficulties in handling what are known as the "dynamic aspects". It would therefore seem beneficial to explore the type of decision support system that should be used to improve the organization of cognitive processes in human beings.

There may be, for example, a need to make future projections. Here a SIMULATION MODEL can be useful. This must be clearly distinguished from the mental model we have assumed that John Smith would form of an ecological case, although both types of model have their uses.

Why is it important to extend the development of simulation models for ecosystems? Obviously the results they give are largely governed by the assumptions on which they are based. A long time ago Daniel Botkin stated that "It (the computer model) forces us to see the implications, true or false, wise or foolish, of the assumptions we have made" > (Hall, 1976, page XVI). A model is an abstraction of a system and a system is any phenomenon, whether structural or functional, with at least two separable and interacting components. Models of ecosystems, therefore, are simpler than real ecosystems. They can be used for predicting behaviour or complicated aspects of reality which are little understood. Since the inter-relationship of business, economic, social and ecological considerations is often very complicated, simulation models have a useful function. Their use should not be viewed as reductionism in the negative sense.

Models allow the application of scientific method to solve complex environmental problems. A possible structuring of the problem solving process involves four steps:

1) Developing a strategy for model building
2) Model building itself

3) The use of the model for analytical purposes
4) Synthesis on the basis of the model

The value of systems theory for many applications in environmental modelling for the solution of complex problems has been sufficiently demonstrated.

References:

(1) Begon, M., Harper, J.L., Townsend. C.R.: Ökologie - Individuen, Populationen und Lebensgemeinschaften (translated from English by Dieter Schroeder and Beate Hülsen). Basel Boston Berlin: Birkhäuser Verlag 1991.

(2) Der Rat von Sachverständigen für Umweltfragen: Umweltgutachten: Für eine dauerhaft-umweltgerechte Entwicklung. Stuttgart: Methler-Poeschel Verlag 1994.

(3) Doerner, D., Kreuzig, H.W., Reither, F., Staeudel, Th. (Eds.): Vom Umgang mit Unbestimmtheit und Komplexität. Bern, Stuttgart, Wien: Hans Huber Verlag 1981.

(4) Doerner, D., Wearing, A.J.: Complex problem-solving: Towards a (computer-simulated) theory (Unpublished manuscript, version 3.0 of 16.3.1994).

(5) Hall, C.A.S.: Ecosystem modelling in theory and practice: An introduction with case histories. New York, London, Sydney, Toronto: John Wiley & Sons 1976.

(6) Lehn, K.van: Problem solving and cognition. skill acquisition In: Posner, M.I.: Foundations of Cognitive Science. Boston: MIT Press 1989.

(7) Patherton, D.P., Borne, P.: Concise Encyclopaedia of Modelling and Simulation. Oxford, New York, Seoul, Tokyo 1990.

(8) Richter, O., Soendegerath, D.: Parameter estimation in ecology. The link between data and models. Weinheim: VCH 1990.

(9) Simon, H.A., Kaplan, C.A.: Foundations of Cognitive Science, In: Posner, M.I.: Foundations of Cognitive Science. Boston: MIT Press 1989.

(10) Stuhler, E.A.: Ökologie und Umweltschutz - Denken, Planen, Entscheiden and Handeln bei der Umsetzung der umweltpolitischen Ziele der Bundesregierung. Working Paper. Freising: Munich Technical University 1992.

Heiner Benking
FAW Research Institute for Applied Processing
at the University Ulm

Visual Access and Assimilation Strategies to Prestructure Bodies of Environmental Knowledge. Proposals and Lessons Learned

Abstract

Environmental problems at all levels, from local to global, are increasingly complex and multidimensional. Effective solutions to such problems often must be interdisciplinary, intersectorial, and international in scope and able to handle management information which in many cases is not harmonized. Numerous barriers to integrating such diverse data exist. Including contextual and application oriented information helps to identify and review information and develop more appropriate retrieval strategies.

The paper reviews design proposals for multi-lingual meta database development and embryonic concepts to access to and navigate in a knowledge map, or better knowledge panorama. The concept can be seen analoguis to optical panning and zooming with microscope and telescope in a scale independent imaginary continuum. In addition, knowledge organization and management is supported by archiving, representation, linking and access techniques, making use of advanced information systems technology. A coarse three-dimensional knowledge-map is proposed, an organization chart or layout, which is linked to geographical co-ordinates and can be referenced to other repositories. The design of the of ECOCUBE allows a coarse but unobstructed view, a picture of the world, in which areas of expertise, bodies of knowledge and observations can be integrated.

Interface design theory and discussions on mental models and cognitive viewpoints form the theoretical aspect of the paper. Practical considerations range from knowledge organization in the field of the environment to the design of a world-view which is consistent, comprehensive, and allows detailing with variable foci and theme compositions.

The objective is to counteract the missing perception of time and magnitudes and address terminological bias by increasing awareness about scales, proportions, and consequences through conceptualization and imagination. Applications in education, harmonization, and resource management are especially suited to detail and improve the concept and test its acceptance and usefulness.

Ulla Pinborg
National Forest and Nature Agency, Copenhagen

Towards a Catalogue of Datasources
for the European Environment Agency

Abstract: The overall task of the European Environment Agency (EEA) will be to produce reliable information for implementation and development of the European environment policy. EEA will coordinate a European information network of institutions in the Member states. A catalogue and a network of institutions with information on environmental data sources (institutions, activities, databases and measurement stations) and a general multilingual environmental thesaurus will be two important EEA tools.

1.   Dealing with environmental data and information calls for harmonization and development of new tools

Awareness that the state of the environment is dependant on the interrelations between natural resources and processes and on the pressures imposed by society, while society is dependant for livelyhood, health and security on the state of the environment is growing in all nations.

Work with environmental issues can only in very few cases be based on own, stand-alone data. Identifying, accessing and using extraneous data is a fundamental part of work with environmental matters. During the past three decades an abundancy of environmental and socio-economic data has been collected and stored. During the same period the technical possibilities for data capture, storage, management and dissemination have changed from being a limiting factor to giving virtually limitless possibilities.

But there are central problems to be overcome for users of this seeming wealth of data and of technical possibilities. So far the majority of data activities have been performed with little or no coordination to other activities, because aims have been limited, and because data interchanges between projects have been few. Now this is changing.

All nations deal with these problems of identification and coordination of widely spread data and black spots in topics, time and space and use large resources in doing so. There are in nearly all countries and in international organisations now integrated data collection and environmental statistic activities as well as activities in standardisation of methods and terminology, and most nations and international organisations work at identifying and giving access to their sources of data.

It is an encouragement towards further collaboration and coordination, that though the work and the solutions vary widely, the basic types of information have large similarities.

2. The European Environment Agency tasks

Within the European Union the European Environment Agency has the overall tasks to produce objective, reliable and comparable information on the environment for those concerned with development and implementation of European environment policy. EEA must provide EU and its Member states with the information necessary for framing and implementing environmental policies and enable the European Commission to carry out its tasks in the field of environmental policy and legislation. But EEA must also ensure a broad dissemination of reliable information to the public.

The European Environment Agency EEA is coming into force in the second half of 1994. The establishment of the EEA is based on the EC Council Regulation No 1210/90 from May 7. 1990. The work has to follow and support the aims of the Council Resolution No 138 of February 1.1993 on a Community programme of policy and action in relation to the environment and sustainable development.

To achieve the aims of the Regulation the EEA work must be shared and decentralized in an information and observation network including the EEA. National Focal Points and the main national component elements of national environmental information networks and European Topic Centres. EEA will also collaborate with a number of international institutions. The work and the collaboration are described in the EEA workprogramme. The first multiannual workprogramme 1994-1999 and the annual workprogramme mid-1994-1995 deal with the first 5 year work-period and specify the tasks and priorities.

In short

* the Member states appoint and describe institutions to partake in the information network
* actual data collection, management and dissemination will be performed under separate projects under the workprogramme involving relevant parts of the national networks and European Topic Centres. A major part of the data will be data already collected or treated by the Member states, such as regulation data
* EEA and the Member states collaborate on establishing and maintaining a physical electronic network for data and message interchange and on guidelines for interchange
* common terminologies and use of identical definitions will be developed to enable collaboration and integration between functions and projects
* information about sources of environmental data and information will be coordinated to allow contact and multiple use of data and information

The data to be handled by the EEA and its network will for the major part originate within Member States as part of national activities or as national parts of international activites. The decentralized function and the simultaneous call for comparability and consistency demand a well-directed coordination.

A number of functions and tools of different character has to be developed or set up to deal with this. Functions and tools to handle specific topic data are based on sectorial approaches. while others are service functions and tools of horizontal nature to deal with technical use of data, with identification of datasources and integration and standardisation of data.

3.   The EEA Catalogue of Datasources project, EEA/CDS

EEA will have an ongoing need to be informed about institutions managing environmental information both as a basis for the EEA Regulation network of institutions and as a basis for the major part of the topic specfic projects. A continuous problem for EEA and its partners will thus concern rapid and good identification and access to sources of environmental data and information in the Member states and international organisations. This need for meta-information is at the same time also a general public need among NGOs. consultants and researchers.

Most Member states have one or more activities concerning identification of sources of environmental data and information. and several countries are developping CDS database systems to cope with the information. Most Member state systems are national. some regional. Several international or large organisations also have CDS systems. There are even a few open commercial initiatives. - Only few of these activities are of a general nature and of several years standing (the Netherlands. France). Most are very recent (France. Italy. Spain. Germany. Austria. United Kingdom) or only just under development. and most are limited in scope to specific environmental topics (water. air. nature. forest etc.), specific functions (research. administration etc.) or specific datasources (institutions. projects and activities. databases. maps etc.).

The need for and use of datasource information varies widely according to the purpose of the use, and the definition of the term meta-information has many facets. -   Many users have initial needs only for simple contact information to institutions as sources of information. But most users want this plus a condensed description of which data and information activities exist and of the range and coverage of and the access to data and information treated in order to evaluate the usefullness of the data for their particular needs. A small. but growing number of users want direct access to the data in databases.

The EEA service function to deal with this will be the EEA Catalogue of Datasources (EEA/CDS) project.

4.   The EEA/CDS aim and organisation

The aim of the EEA/CDS project will be dual :

*   inspiration and harmonization of initiatives in Member states to ensure wider access to and use of data
*   collection of datasource information to be used by EEA and its partners in concrete projects for orientation in the market of data and information

The EEA/CDS work will have the form of a gradual process and must of necessity be run simultaneously on two interconnected tracks :

*   development of the EEA/CDS own central system to serve the EEA central work
*   collaboration on access and harmonization of concepts, contents and functionalities of collaborating systems

The EEA/CDS project organisation will consist of a network of collaborating national CDS activities and a central EEA activity. The national activities form the main basis for obtaining meta-information for EEA directly, and actual collaboration may take the form of formal agreements with general CDS activities or be part of topic projects, different as they will be.

The EEA/CDS work is foreseen to be constituted as a horizontal EEA project with an EEA/CDS group from Member states and a small number of other relevant partners to act as advisors.

5. Catalogue of datasource concepts

The first CDS systems were fashioned along the lines of extended telephone directories or library document catalogues, with a very firm structure and relying on heavy use of keywords. Most systems of this type were built to fit rigid paper forms. Recent systems contain telephone directory parts, keyword parts and full texts, the full text parts becoming more and more extended and free, using new full text managemnet techniques and moving from paper form to digital form. In this process sometimes discarding the userfriendly simplicity of the older systems, but gaining in freedom and expressivity.

Some of the systems are developped to fit a very decentral capture of detailed meta infor-mation close to the actual data collection and to measurement stations. They may be seen as integral parts of measurement or monitoring activities. Such bottom-up systems aim to depict in some detail, what is actually going on at specific sites. Such systems are not from the start directed towards information on institutions or on data products. Structured keyword lists are necessary tools. The definition of meta-information in such systems is detailed and stops short of the actual data. Decentral updating is simple, central updating is difficult. Collaboration with other systems calls for aggregation of data and harmonized definitions of terminology. The bottom-up approach may fit such EEA projects as the more detailed topic projects, where information about parameters and geographical distribution of stations and measuring network coverage is essential.

Most systems to date are top-down systems aimed at overall orientation of actors implica-ted in data and information activites. The information is condensed or aggregated. Such systems are low in level of details and in number of keywords used. The form is close to traditional directories. Decentral updating is simple, central updating is also simple, but timeconsuming. Parts of such systems can easily be coordinated, but definitions and topic categories may be somewhat unclear.

Other systems of newer concepts are hybrids between the two, extending the computer technical possibilities to be used also in the descriptive concepts, operating with combinations of fixed structures and full texts, with possibilites to affix other documents and to relate in a flexible way from entity to entity. These systems combine the flexible word processing and document-orientated work-day conditions of many CDS users with the more structurally demanding database use and allow a multimedial approach. Meta data collection and use of forms for such systems can be done in a variety of ways with different forms and media and thus calls for modular concepts and good guidelines. Collaboration between such slightly "anarchistic" systems may not immediately be seen as easy. But harmonization in description of structures and contents to be interchanged and development of common terminological tools move the collaboration and coordination problems from being connected to the physically collaborating systems to the data exhanged. This development is common to all data work at present.

The national or other CDS systems so far developped have normally not been customized in coordination with other systems. The general concepts therefore vary considerably, and so do the specific field structure and data content of the system databases. But there are also large similarities, because of parallellity in content types and use. These similarities arise from the nature of the environmental data content and management and form an encouraging basis for collaboration and harmonization. -   Within the EEA/CDS project there will be collaboration with systems of all these types and with all these types of structures and contents.

6.   The EEA/CDS concept

The EEA/CDS concept must rest on the specific needs within the EEA for its dayly work and for the EEA projects (including general information of the public) and on the concepts of the collaborators.

The preparation for the EEA/CDS has consisted in getting inspiration from existing CDS initiatives and in analysis of common needs for future uses of the EEA/CDS.

The concepts of datasources in existing systems fall within 4 types :

*   institution or expert person responsible for information or data
*   data or information activities such as monitoring or publishing programmes and
         projects
*   data products , i.e data or information stores or containers such as databases. maps.
         documents
*   measurement stations or sites

Some systems contain all 4 types. some only 1 or 2 and with greatly varying level of detail on the types. The meta-information content is expressed in a varying number and arrangement of informative entities and items and at different levebof detail. This is where the greatest dissimilarities between systems are found. Often the definitions of informative items also vary or are vague.

For the EEA use there is a need for meta-information on all 4 types of datasources and for identification of relations between many of them.

Nevertheless 8 basic types of major informative entities can be identified :

*   institution
*   person
*   postal addresses
*   tele communication addresses
*   activity
*   product
*   data
*   station

The 8 entities occur in varying patterns in meta-information on all 4 datasource types. They can be used as bricks in a modular CDS system for the EEA and as conceptual modules for harmonization and interchange of information between partners.

The 8 entities found in CDS systems can be analyzed to contain ex- or implicitly informative items of similar structure.

For the EEA/CDS the 8 entities must cover all or most of the following 10 informative entities :

*   system internal references
*   name
*   type
*   function
*   relations to relevant other informative entities
*   topic/parameters
*   temporal aspects
*   spatial/geographic aspects
*   references to further documentation
*   reference to respondent to CDS system

The suggested EEA/CDS will thus consist of

*   4 columns to be treated both in unison and separate with information on
        institutions, activities, products and stations
*_   for all columns a modular structure based on the 8 entities and the 10 types of
        informative items
*   combination of use of fixed fields, structured and free keywords, full texts and
        affixed documents

For each particular case (specific catalogue) it will allow select use of relevant modules only and opening for insertion of other modules. The number of relevant informative items and of actual information content will vary between the 4 columns, but the overall structuration of information will remain.

7.   The multilingual challenge

The EEA/CDS working reality will be multilingual. The users may in principle want to use any own language. The collaborating institutions and CDS systems are directed towards national use and collect and present their meta-information in the national language(s), though a few begin to have part information in English. The international CDS systems of greatest immediate EEA/CDS relevance are based on English. EEA will get meta-data in many linguistic versions and have to deal with this in different ways (agreement on bilingual input, translation, coding, keywording, full text search).

To deal with all meta-information, to present it and allow access to it in any of the EEA languages is a natural, but removed ultimate goal. To deal only in English will be satisfactory to few.

From preliminary collaboration and use of the first modules so far there emerges a possible picture of :

*	database technical language	often based on English (SQL,CCL etc)
*	database presentation screens	first version in English,
	for EEA/CDS open database use	later more languages
*	guidelines	major languages
		finally perhaps all languages
*	technical handbook	English, perhaps other major languages
*	meta-data	input data texts after agreements in national language and English
*	use of authority terms	any language as input, user choice in output
*	use of keywords	standard keyword input in any standard list language, free keyword input in national language or English
*	publications	dependant on meta-information and needs. Short versions not based on texts in all languages. Versions based on texts dependant on text language

The important issue is to keep the national activities in the national languages national, and at the same time to obtain from existing activities good information with the smallest possible extra effort in languages of use for many. - The immediate tools to ensure wide multilingual use are the standard term lists used for authority terms and keywords or as basis for text searches. They have to be developed on a multilingual basis.

AUTHORITY LISTS are short lists of system specific standard terms or codes (often for fixed tick-off fields). They will be developed in coordination with the advisory group.

STANDARD KEYWORDS may be either of a general or of a more specialised nature. For use in the EEA/CDS and nationally for most keywords of a general nature, a General Multilingual Environmental Thesaurus is already under development in 6 languages (Dutch,

Italian, English, German, Spanish, French), and more languages will follow.

SPECIALISED KEYWORDS must either be free keywords in any language or preferrably in English or in a scientifically widely used term or code (chemical substances, living organisms). For some aspects the special standard term lists must be developed on a national basis such as for geographic names.For other topics lists can be developed/adopted in international collaboration between general term institutions and scientfic/administrative expert institutions (chemical substances, living organisms, soils, wastes etc.).   -   Free general keywords may be in any language, but preferrably bilingual.

By means of extensive full text search facilities for database versions of the EEA/CDS (nearly) all terms in any language can principally be located, but not structured and aggregated. For each printed version a language choice must be made dependant on the users and the meta data in existence.

8.   Use and products from of EEA/CDS

The users of meta-information will normally address the national CDS system and responsible for national data first. The users of the EEA/CDS will be the EEA itself for dayly and project orientated use, the European Commission and international partners, but also users needing meta-information beyond national boundaries. The national systems may have any form or functionality from being simple information in a responsible institution via simple directories to advanced or special reports and databases.

The EEA/CDS must be developed in a flexible way allowing for several forms and levels of input. The EEA/CDS is foreseen to hold a central master database and to issue reports or database versions with varying functionalities from this.   -   A number of paper forms must be developed, but for the main inputs from collaborating partners computer-forms should be encouraged or use of interchange formats for downloads from existing databases.

An important aspect for all EEA work is the geographic dimension, and both the central database and the printed or electronic versions must have easy-to-use geographic functionalities. The geographic functionalities are wholly dependant on the validity of the geolocation information stored and retrievable. Therefore all geolocated input data are requested to be followed by at least a geo-referencing code to administrative units such as region or commune, and use of geographic coordinates are encouraged. But maps may also be added as annexed documents for visual inspection and representation. Geolocational aspects are so far not very well developed in most CDS systems.

The first version of EEA/CDS institution formats on paper and diskettes have in 1994 been tested on descriptions of national environmental institutions, while the form concepts have been developed for surface water and air quality monitoring activities and on species databases. The first product will be a survey of the EEA information network institutions.

**   The author has prepared this paper as part of a consultancy study for the EEA Task Force in Bruxelles to help prepare the workprogramme for the EEA. The views expressed are those of the author.

Thomas Schütz, Helmut Lessing
Niedersächsisches Umweltministerium, Hannover

The Umwelt-Datenkatalog (UDK) of Lower Saxony Structure and Functionality

Abstract

A current problem in information systems is the ever growing data quantities that make it more and more difficult to keep track of collected and stored data in every field of public interest, including the administration of environmental concerns. In November of 1990 the cabinet of Lower Saxony therefore resolved to develop and to install a catalogue of data sources concerning environmental interests and tasks. This catalogue was called **Environmental Data Directory** or **Umwelt-Datenkatalog (UDK)** [(1) - (7)].

Begining the implementation of this solution in 1991, a first information management system based on electronic data processing was developed. At the end of 1992 the version 1.3 was completed and first tests were begun with a few chosen administrative authorities. Up to now ten states of Germany cooperate in coordination and financing the further development of the software UDK. They are also standardizing the description of meta-data.

In 1994 the software became more powerful, the data model was enlarged and the mechanisms of search and printing and the graphical user interface were improved in the version 1.5. Furthermore, Germany and the Republic of Austria signed an agreement on common use and development. Germany is responsible for the development of the software UDK, which is delegated to the Ministry of Environment of Lower Saxony, whereas Austria is concerned with the development of a thesaurus and its integration into the UDK.

The main advantages of such a data directory as the UDK are the following:

- the most complete survey possible of data concerning the environment, collected by or stored at the administrative authorities
- precise description of the data quality
- access to the original data by network request
- supraregional standardization of the description of data sources, the so-called meta-data

In the UDK a database is described in terms of meta-data. This description is called the UDK-object. All UDK-objects are stored in a table of the database. They are connected to the nodes of one or more trees that lead to a hierarchical structure of the UDK-objects. The only information which the nodes of a tree contain is its name and its position in the tree marked by a specific number. There are two different kinds of trees. All the nodes of a primary tree point to exactly one UDK-object and every UDK-object is represented by a node of a primary tree. The secondary trees consist of nodes that need not point to a UDK-object, e.g. they are default nodes. The Grunddatenkatalog (GDK) of Germany is an example for a secondary tree. Objects connected to such nodes are also referred to by a node of a primary tree. Besides, the software UDK can manage several primary trees so that it is possible to include the primary trees of several states and countries.

References

(1) Lessing, H: Umweltinformationssysteme - Anforderungen und Möglichkeiten am Beispiel Niedersachsens. In: Jaeschke, A., Geiger, W.. Page, B. (Eds.): Informatik im Umweltschutz 4. Symposium, Karlsruhe 1989.

(2) Lessing, H., Weiland, H.-U.: Der Umwelt-Datenkatalog Niedersachsens. In Pillman, W., Jaeschke, A. (Eds.): Informatik für Umweltschutz, 5. Symposium, Wien 1990.

(3) Lessing, H., Schmalz, R.: Der Umwelt-Datenkatalog Niedersachsens. In: Engel, A. (Eds.): Umweltinformationssysteme in der öffentlicher. Verwaltung. Heidelberg 1994.

(4) Schütz, T., Lessing, H.: Der Umwelt-Datenkatalog Niedersachsens. In: Hutzinger, O. (Ed.): ECOINFORMA'92, 2.Internationale Tagung, Bayreuth 1992.

(5) Schütz, T., Lessing, H.: Metainformation von Umwelt- Datenobjekten - Zum Datenmodell des Unwelt-Datenkatalogs Niedersachsens. In: Jaeschke, A.. Kämpke, T., Page, B., Radermacher, F.J.(Eds.): Informatik für den Umweltschutz, 7. Symposium, Ulm 1993.

(6) Schütz, T: Der Umwelt-Datenkatalog Niedersachsens - Informationsmanagement für den Umweltschutz. In: Denzer, R., Geiger, W.. Güttler, R. (Eds.): Integration von Umweltdaten, 1. Workshop, Schloss Dagstuhl 1993.

(7) Schütz, T., Wagner, H.: Metadatenklassen. In: Denzer, R., Geiger, W., Güttler, R. (Eds.): Integration von Umweltdaten, 2. Workshop, Schloss Dagstuhl. To appear 1994

Wolf-Fritz Riekert
FAW Ulm (Research Institute for Applied Knowledge Processing)
POB 2060, 89010 Ulm, Germany; E-Mail riekert@faw.uni-ulm.de

Management of Data and Services
for Environmental Applications

Abstract: In recent years, systems for processing environmental information have been evolving from research and development systems to practical applications. Many of these systems already support environmental activities of the public sector. Most of these systems, however, were developed as island solutions. The integration of such heterogeneous systems within one distributed environmental information system, is a big challenge of today. It requires new methods to navigate to the services and data offered by the components of such a system. Metadata (such as data and service dictionaries) and codata (such as spatiotemporal references) must be provided in order support the proper usage of these data and services. Systems integration techniques are required in order to overcome the heterogeneity. An approach to fulfill these requirements is shown by examples from recent work at FAW which has been conducted in the context of the Environmental Information System Baden-Württemberg.

1. Environmental Data

Environmental data differ from traditional data in many respects. Often the data are related to a certain location and time interval. Measurement data that are acquired for discrete locations and points in time belong to a special category; these data need to be interpolated if continuous profiles are required (e.g., air pollution data or elevation models). A special case is image data (e.g., satellite sensor data), which typically consist of very large raster data sets. NASA experts often handle terabytes of satellite data; this is more than two orders of magnitude greater than the amount of data managed today in large international financial transaction systems. This requires new storage models for environmental databases, e.g., the use of automatic tape archives and CD ROM jukeboxes as tertiary storage medium.

The most demanding problem concerns the problem of deriving information of interest from existing data. In the ideal case, an environmental database contains all necessary data which are needed to derive the information requested by the user. Apart from the raw environmental data, however, this requires the availability of additional data. These additional data are often refered to as *metainformation* or *metadata* (6). *Codata* is another term which is also used in this context. In this paper, we try to differentiate between metadata and codata. Metadata provide abstract descriptions of the data structures and data formats used in the underlying system, whereas codata include additional instance-specific data about location, time, precision, and revision dates of the data under consideration. Location and time are very important kinds of metadata which are discussed in more detail in the next section.

Metadata and codata are missing today in most existing environmental information processing systems. Often the only data source consists of a magnetic tape containing the raw data (e.g., image data or measurement data) to be analyzed. All of the additional information which is necessary for the correct interpretation of the raw data is implicitly coded in the application programs or must be contributed by the user. Such a tape may become completely useless after a change of the personnel or of the data processing software. Therefore an effective management of metadata and codata is of crucial importance for environmental information processing systems.

2. Spatio-temporal Aspects

Environmental data often describe environmental objects with a spatial and temporal extent. These objects, also known as geographic objects, possess a geometry consisting of point-form (0D), linear (1D), flat (2D), or solid (3D) features. Often geographic objects also have a lifetime which is given by their dates of construction and destruction. During the lifetime of an environmental object, its attributes may change. That is, the values of the attributes are only valid during a certain time interval, and a given attribute (e.g., the population of a city or the land use of a parcel) may possess many possible values during the lifetime of the object. A special case is the change of the geometry of a geographic object such as the growth of a city or the shrinking of a lake or a forest.

Spatio-temporal data are important examples of codata in the environmental domain. The management of spatio-temporal data is a special challenge to applied computer scientists. Abstract data types are required for representing concepts such as the partonomy, the topology, and the spatio-temporal extension of geographic objects as well as the thematic information associated with these objects, such as alpha-numeric attributes and relationships to other geographic objects. An important task is also the (carto-)graphic presentation of these objects. In most cases, non-standard data types are required for representing these kinds of information and, therefore, a trend towards systems which allow the definition of such data types, such as extended relational or object-oriented databases (1), can be recognized in current developments (3).

The special nature of environmental data also requires a query language with special characteristics. Apart from typical SQL-like questions, the forthcoming object-oriented database management systems (2) also allow navigational queries and queries that include user-defined predicates. In particular the latter allows the usage of spatio-temporal predicates in queries. The optimization of spatio-temporal query-processing is a scientifically demanding problem which concerns both storage models and indexing techniques for multi-dimensional data. The integration of these techniques with existing databases is a hard problem which will still require major research activities in the future.

3. Overcoming the Heterogeneity

The harmonization of environmental information at national, European and worldwide levels is of central importance for gaining a reliable description of the environmental situation and, at the same time, is a basic requirement for any reporting system in this context. These requirements, however, are confronted with the existing heterogeneity of hardware and software environments, of database systems, of method bases, of network technology, and of various computer languages.

The task of overcoming heterogeneity also requires the availability of meta-information. Doubtless the development and promotion of standards is of particular importance in this respect. Experience shows, however, that we will still have to cope with competing standards in the future. In addition, technological advances will always produce new heterogeneity problems and will require strategies for migrating the software towards new solutions. Meta-information about data and methods can be used in order to do the necessary translations between different systems. In the public sector, environmental metainformation systems, also known - with a reduced scope - as *environmental data directories* (8), are currently being developed to overcome the prevailing lack of metainformation in existing environmental information systems.

4. Environmental Information Management in the Public Sector

In recent years, environmental protection as an objective of public activities has reached quite a high standard in Europe, particularly in Germany. Communal tasks, for example, include land use planning, approving compliance with environmental standards, management of hazardous waste, and water and energy management for public facilities. Several states and regions of the European Union have already installed effective environmental information systems and powerful sensor networks.

Official systems such as the Environmental Information System (German acronym: UIS), Baden-Württemberg (5) support environmental tasks at various levels: Decision Support Systems are provided for the high-level environmental management, reporting and planning systems are available for middle management, while basic components support the acquisition and management of specific environmental data at the operational level. In addition, interdisciplinary information systems are being used in public environmental management; these systems are not restricted to environmental tasks such as the topographic or cadastral information systems of the surveying offices.

UIS is the organizational, informational, and task-oriented framework for the supply of environmental data and for the processing of both department-specific and interdisciplinary tasks in the environmental domain of the State administration. UIS consists of a large number of components that are implemented on various hardware and software platforms and are operated by various departments at distributed locations. In this context, the INTEGRAL project, which is presented in the following section, aims at a user-friendly and economical way of accessing the functionalities of these distributed system components.

5. INTEGRAL

The INTEGRAL (Integration of Heterogeneous Components of the Environmental Information System of Baden-Württemberg) project has been conducted since 1993 at FAW under commission of the Environment Ministry, Baden Württemberg. The central goal of this project is to continuously increase the integration of UIS components and to improve the ease of use of these components from remote sides (7). The networking concepts available in open systems are a promising option for the necessary integration, because these systems provide highly effective mechanisms for sharing data and functionality within a computer network. Since all commercial operating systems support these communication standards to a large extent, this approach does not severely restrict the applicability of the INTEGRAL concept to dedicated computer systems. In INTEGRAL, the FAW is working on communication interfaces oriented that are oriented towards a client/server model which is typical for modern system architectures. This kind of model is also the target of FAW´s software engineering strategy in general.

At the highest level of abstraction, an interactive interface based on X Window allows the functionality offered by different systems to be shared at the screen level. At a lower level of abstraction, a service interface developed by the Institute for Nuclear Energetics and Energy

Systems (IKE), University of Stuttgart, is provided. This interface makes the functionalities from existing programs available in the form of self-contained services.

The access to these functionalities is based on a hypertext system. The management of these functionalities is supported by meta-information. In this way, preferences of the users may be taken into account. For example, urgent messages may be transported as rapidly as possible, whereas large simulation jobs may be computed at inexpensive rates during the night. Beyond integrating computer systems, INTEGRAL also connects people by empowering them to work cooperatively and to communicate efficiently by means of electronic media.

6. Conclusion

The management of data and services for environmental applications using codata and metadata is of crucial importance for environmental information systems. The INTEGRAL approach shows that the availability of data and services in environmental information systems will be enhanced and standardized by the techniques proposed. The advantage arises from the simplified, uniform, and more user-friendly availability of services offered via the network. Such services can be shared from client workstations all over the network. The "server-side" location of the programs and data that provide the services and information reduces hardware, software, and maintenance requirements.

References

(1) Special Section: Next Generation Database Systems. *Communications of the ACM*. 34(10):30-120, October 1991.

(2) Cattel, R.G.G.: *The Object Database Standard: ODMG-93*. Morgan Kaufmann, 1994.

(3) Günther, O.; Riekert, W.-F.: The Design of GODOT: An Object-Oriented Geographic Information System. *IEEE Data Engineering Bulletin* 16(3), September, 1993.

(4) Jaeschke, A.; Kämpke, T.; Page, B.; Radermacher, F.J.: *Informatik für den Umweltschutz*. Springer-Verlag, Berlin – Heidelberg – New York, 1993.

(5) Mayer-Föll, R.: Das Umweltinformationssystem Baden-Württemberg; Zielsetzung und Stand der Realisierung. In: (4), pp. 313-337.

(6) Radermacher, F.J.: The Importance of Metaknowledge for Environmental Information Systems. In: Günther, O.; Schek, H.-J. (eds.): *Large Spatial Databases*. Proceedings. Lecture Notes in Computer Science 525, pp. 35-44. Springer, Berlin – Heidelberg – New York, 1991.

(7) Riekert, W.-F.; Henning, I.; Schmidt, F.: Integration von heterogenen Komponenten des Umweltinformationssystems (UIS) Baden-Württemberg. In: 2. *Workshop „Integration von Umweltdaten"*, KfK 5314, Kernforschungszentrum Karlsruhe, 1994.

(8) Schütz, T.; Lessing, H.: Metainformationen von Umwelt-Datenobjekten – Zum Datenmodell des Umwelt-Datenkataloges Niedersachsen. In: (4), pp. 19-28.

Konrad Zirm
Austrian Federal Ministry for Environment, Youth and Family

Environmental Information in Austria
Advances in Environmental Meta-Information Systems

Abstract

Environment data and respective environment information are being collected by state-
or environmental agencies and by international organizations and are made available in
the form of "Environment Reports". In addition to this, environment institutions,
environment research programs, conferences and databases, data networks and
mailboxes facilitate integration and exchange of environment data.

It becomes increasingly obvious to display the sources of such data and to make
transparent the generation of its information. Because of this, meta-databases on
environment data are being conceived. It can be noticed that in Germany the federal
states increasingly agree in a uniform presentation of meta-information on environment-
relevant data on the basis of the Environment Data Catalogue (UDK) of Low Saxonia
(Niedersachsen)

To have access to information from all activities of administration has been legalized in
a number of industrial countries. Examples are Finland, Italy, Canada and Sweden. In
the USA, the "Freedom of Information Act" already dates back to the year 1967. The
guidelines of the European Community on "The Free Access to Information on
Environment" (90/313/EWG) has been transferred into national law by the member
states. In Austria, too, a federal law on the access to information on the environment
(Environment Information Law) was released, July 1993. In this law the etablishment of
an Environment Data Catalogue was made obligatory.

The meta-database developments are attractive from the point-of-view of an
internationally standardized access to environment data. It would be desirable to create
an international environment data catalogue of which the costs for its further
development could be shared. In Austria, for example, it is planned to develop a
multilingual thesaurus as a supplement to the UDK (Federal Ministry for Environment,
Youth and Family, Vienna). The European Environment Agency (EEA) supported the
establishment of a tool for the cataloguing and access of environment information.

Zdravko Krakar
Institute for Information Technologies, Zagreb, Croatia
Nenad Mikulic
Ministry of Civil Engineering and Environmental Protection
Zagreb, Croatia

A Metastructure of the Environmental Management Information System in the Republic of Croatia

Abstract: Respecting the significance of the national environmental management issue, the Ministry of Civil Engineering and Environmental Protection of the Republic of Croatia has initiated a programme the intention of which is to provide guidelines for establishing a methodology of environmental management and a proper information system serving this specific function at the national level. The paper presents the results obtained so far. On the basis of the World Resources Institute perceptions, the recommendations of Agenda 21, and the analyses of some international experiences, as well as knowledge of the situation in Croatia, general principles and instruments necessary for the establishment of a rational national environmental management system have been generated. The article represents the Croatian approach to solving such complex information system goals, contents structure and the way of its creation and building.

1. Introduction

Social development of the Republic of Croatia, as well as that of other countries, is based on two apparently contradictory goals:

(1)   economic growth through the increase of production, and

(2)   improvement of the environmental quality by decreasing the present degree of its pollution and degradation.

At this day and age, managing national environment represents an extremely responsible task. Intensive exploitation of environment is based on the concept that environment and its resources are inexhaustible. Demographic growth, urbanization, industrialization and development of technology have led to serious disturbances in the ecosystem, but also to new conceptions. The principles of classical industrial development lead to an environmental disaster. On the other hand, there is evidence enough of the threat represented by the menacing environmental dictatorship. To find the middle path seems like the only reasonable solution. However, it still differs very much from the so far valid and known ways of development. It is the obligation of each and every state to find solutions adjusted to its own characteristic conditions. According to modern conceptions, environment has at least nine dimensions: economy, technology, population, natural resources, environmental pollution, urbanization, space, politics and socio-cultural relations. How to plan and implement national development based on the interaction of these factors represents a crucial issue for all the countries of the world. Solutions are sought in appropriate concepts.

The possible ways of managing environment point to 5 possible concepts: frontal management, resource management, sustainable development, selective environmental approach and deep environmentalism.

The paradigm of sustainable development today represents a topical issue worldwide. Already in the fundamental document that, back in 1987, introduced it as the modern developmental strategy, it has been stated that it is not a state of harmony in society, but rather the process of change in which resource use, directing investments, technological orientation, as well as changes in society's institutions, become compatible with both the present and the future needs.

2. Review of the Present State in the Republic of Croatia

The final document of the World Conference on Environment and Development, Agenda 21, imposes on the Governments of the countries-signing parties the obligation of adopting a national sustainable development strategy.

The Republic of Croatia has not yet adopted its national strategy in compliance with the above document. However, in order to illustrate the present state in the country in this respect, we may make use of the Recommendations adopted in the course of 1993 by the IUCN - the World Conservation Union, through its Commission for Environmental Strategies and Planning (CESP).

The said Recommendations require the following: one integral and not several sectorial strategies, the sense of need, support by political authorities, active involvement of key participants in the strategy implementation, favorable social climate, the necessary financial means and the strategy implementation control mechanism. With regard to the above recommendations, the situation in the Republic of Croatia is as follows: national strategy has not been adopted yet, but it is being elaborated through the aforementioned programme. The need for it and its purpose are felt in many segments of the society, but the respective wholes are still not sufficiently connected. The political support does not amount to much at the moment. Croatia has only just come out of the patriotic war, considerable parts of its national territory are still under occupation, and the new state organization is just being created. Under such circumstances, there are other national priorities. However, regardless of such conditions, the programme by the Ministry of Civil Engineering and Environmental Protection has been joined by most other ministries, and a large number of other institutions. Many concrete projects - parts of this programme - were created under real conditions of war.

The general state of neither war nor peace, as well as the existing economic situation, have been responsible for the lack of financial means, but this nevertheless does not represent a major hindrance to the national programme development.

3. Method of Implementation

The implementation of such a national programme is to proceed in several phases. It has been observed that we may distinguish at least between 5 such phases: phase of unconnected activities, initialization, constituting, immediate implementation and maturity phase.

The initialization phase was started three years ago by observing the need for it, and then animating other ministries, institutes, project and other cooperating offices. It lasted about a year. The constituting phase encompassed the establishment of a Government Commission co-ordinating the entire programme, the making of several pilot projects, but also its methodological and content-related formation. Three principal environmental management domains have been observed: environmental planning, use and protection.

Table 1 shows the contents-related structure of the above basic environmental management domains.

The immediate implementation phase is under way and encompasses 4 developmental directions:
  (1)   Designing the system per priority thematic areas.
  (2)   Establishing an organizational infrastructure and passing the necessary
        laws and other regulations.
  (3)   Initializing concrete regional projects (Danube-valley Catchment
        Area, Lonja Field, Drinking Water for Istria, Oceanographic System,
        Waste Management)
  (4)   Developing the geo-basis.

About 30 projects are currently undergoing the procedure of being adopted by the

Government of the Republic of Croatia. They shall be joined into a single national programme.
The programme's maturity phase is obviously yet to follow.

Planning	Use	Protection
Official topography and cartography	Raw materials, geology	Environmental indicators
Terrain models	Industry	Emission/imission cadastres
Land cadastres	Power supply	Water and water resources
Spatial development plans	Traffic	Air
Physical planning	Agriculture	Soil
Geodetic and spatial system	Forestry	Flora and fauna
Spatial basis of territorial units and settlements	Tourism	Oceanographic system
Networks of watercourses	Fishery	Littoral
	Communal activities	Waste
	Settlements. housing	Noise
		Protected zones of nature
		Cultural Heritage
		Accident states

Table 1: Structure of the National Environmental Management Programme

4. Conclusion

The goal of the National Environmental Management Programme is to shape methodological bases and develop informational infrastructure enabling them to use information and communication technology. Since environment is a strategic resource of each country, and since several state bodies are taking part in the building of an informational infrastructure, it is necessary to join individual projects from this domain into a national programme in order to ensure synergy, define priorities, rationalize the use of resources, coordinate the work on projects and finish them within the envisaged terms.
The projects of building an environmental management information infrastructure and the completion of implemented regulations should be coordinated both in terms of content and that of schedule, in order to ensure consistency, feasibility and efficiency of the entire system, and avoid redundancy in data gathering. This is particularly significant with regard to the present system, because data gathering requires major financial investments, which means that co-ordinated programmes are able to achieve considerable savings.

References:
(1) EECONET (1991): Institute for European Environmental Policy: Towards a European Ecological Network. Arnhem, Netherlands. 80 pp.
(2) OECD (1991): Organization for Economic Cooperation and Development, The State of the Environment, OECD, Paris.
(3) UN (1992): United Nations Conference on Environment and Development, Rio de Janeiro, June 1992: Information for Decision-making, Agenda 21, Section IV, Chapter 40, Doc. A/CONF 151/4 Part IV, United Nations, New York, 65 pp.
(4) IUCN/CESP The World Conservation Union - Commission for Environmental Strategies and Planning, WG on Strategies for Sustainability, National Sustainable Development Strategies, Review Draft, Gland, April 1993, 5+74 pp.

Ivan Duša
Ministry of the Environment of the Slovak Republik
Bratislava

The Environmental Information System in Slovakia

Abstract

The concept of an environmental information system for the Slovak Republic (EIS SR), approved by the Government of the Slovak Republic in May, 1992 defines EIS SR as a tool for the collection, processing, storing and dissemination of environmental information. The EIS SR is oriented towards four main groups of users, namely:

Central Governmental Bodies and Political
Institutions of the Slovak Republik (e.g. ministeries,
　　parliament, etc.)
State District Offices and Local Government
Professional Institutiones (e.g. research
　　institutes, industry, etc.)
Public

The EIS SR can be characterized as a set of information sources, technical (HW), software and communication tools, operational and legislative rules inter-connected by conceptual and methodological approaches used for the implementation of large information systems.

The EIS SR is based on the following principles

EIS SR is an open information system
EIS SR is a distributed information system
EIS SR is GIS-oriented

The next task in the process of establishing of EIS SR is to prepare a comprehensive project of EIS SR. With respect to the extreme complexity of the problem it is necessary to use the experience of countries and companies which have analogous systems realized in the past.

57

Wolf-Dieter Batschi
UMWELTBUNDESAMT (FEDERAL ENVIRONMENTAL AGENCY)
Berlin, Germany

Environmental Thesaurus and Classification of the Umweltbundesamt (Federal Environmental Agency), Berlin

Abstract

The main elements of the environmental thesaurus and classification scheme of the Umweltbundesamt Berlin (Federal Environmental Agency) are described. Their effective application for the environmental literature, the research and development projects and the environmental law databases is shown. The various connections of these tools to international and national activities and projects dealing with environmental terminology (e.g. multilingual general environmental thesaurus for the European Environmental Agency, multilingual environmental thesaurus for the catalogues of datasources in Austria and Germany etc) are given. Further developments and future applications of the environmental thesaurus and classification are presented.

1. Introduction

My paper will deal with the following aspects:
- what are the main fields of application of the thesaurus and classification
- what do one need to know about the history of the thesaurus and classification
- what are the main principles of the structure
- what tools are used for handling the thesaurus and classification
- what are the relations and connections to other activities in the field of environmental terminology
- what are the next steps for the development of the thesaurus and classification.

2. Main areas of application of the environmental thesaurus and classification

The Umweltbundesamt (Federal Environmental Agency) was established as an "independent superior" federal agency in 1974. The main tasks are defined in the act by which the agency was founded. These duties are:
- to provide scientific support to the Federal Ministry for the Environment, Nature Conservation and Nuclear Safety, (in particular: preparation of legal and administrative regulations)
- examination and development of appropriate measures for the protection of the environment
- development and operation of an information system for environmental planning (UMPLIS) and a centralized environmental documentation
- information of the public
- prepare central services and assistance for research activities of the ministry and for the coordination of the environmental research work of the federal authorities.

What happened since that year of establishing the agency?
- Several environmental databases have been designed, developed and are in operation
- the greater part of these databases are open for an online access through the public since 1984 (far ahead of the legal obligation by the EG-directive on free access to environmental information of 1990, which came into force in Germany in July 1994)
- set-up of an infrastructure for answering questions from the public by means of distributing booklets, folders, posters, leaflets etc
- creation of a cooperation with the environmental agencies in the german states (laender), based on the principle of an equivalence of the contributions of the partners.

The databases in operation - according to the above mentioned mandate - are:

ULIDAT / ULIT   Environmental Literature Database
(> 200 000 documents from the german speaking countries)

UFORDAT / UFOR   Environmental Research and Development Projects Database
(> 32 000 r&d-projects and 8 800 institutes)

URDB   Environmental Law Databases
(4 500 court decisions, 11 000 laws (Federal and Laender level), 350 international laws (treaties), 2 500 laws from the European Union)

MONUFAKT   Database on Damages to Monuments caused by Environmental Problems
(> 3 100 items)

These databases (except MONUFAKT) are available online to interested users via several hosts in germany and switzerland.

For the indexing the publications, the research and development projects, the monuments and the description of their damages, repair methods and materials, the laws and decisions of the courts in the above mentioned databases, the environmental thesaurus and classification are used.

3. History

When we started our activities in 1974 we first of all tried to get an overview on:
- existing databases and thesauri or classification schemes
- appropriate software for preparing databases and handling thesauri.

We had to experience that there were a lot of small documentation centers in germany with very different indexing tools and rules for the input of literature. No thesaurus existed for the complete range of environmental questions to be dealt with.
On an international range we had to determine that only an unimportant number of publications from german speaking countries was contained in internationally operating databases (e.g Environment Abstracts, Pollution Abstracts) and no thesauri were used.

Therefore we decided to start databases for environmental literature and research and development projects for the german speaking community, and to develop an environmental thesaurus for the german language together with a classification scheme.

Our first attempt to develop the thesaurus was a more theoretical or scientific approach: The thesaurus was constructed as a tool, coming out of the various lists of index terms used at the different documentation centers. It should become something like a "roof" thesaurus. We spent a lot of money and time doing so, but we had to experience that this approach did not work, because during building up the thesaurus in a theoretically correct way, a lot of index terms were "invented" to complete hierarchy steps. This lead to a mass of terms and brought the thesaurus in a state of inoperability. We had to change our direction and we decided to take as controlled terms for the thesaurus those terms actually used for indexing in the databases , looking also at the free terms which were used for an important number of documents. These terms went throug a checking procedure and after harmonization became controlled index terms of the thesaurus and were brought into a hierarchical ranking, considering only really used terms.

4. Structure of the environmental classification and thesaurus

a) Environmental Classification:

Our classification represents a more conservative view to the environment: the traditional media are used as main groups. These areas reach from air to water, as you can see in figure 1:

Environmental areas

AB	Waste
BO	Soil
CH	Environmental chemicals/pollutants
EN	Environmental aspects of energy and raw materials
GT	Environmental aspects of organisms, viruses and genes, mutated by genetic engineering
LE	Noise/vibrations
LF	Environmental aspects of agriculture and forestry, fishing, nutrition
LU	Air
NL	Nature and landscape / regional development
SR	Radiation
UA	General and interrelated questions
UR	Environmental law
UW	Environmental economics
WA	Water

Fig. 1: Main Areas of the Environmental Classification

For the classes there exists a subclassification, valid for all areas except the groups UA=General and Interrelated Questions and UR=Environmental Law, shown in figure 2:

Subclassification of environmental areas

10   Origin, occurrence and development, nature and dispersion of environmental pollution
20   Effects of environmental pollution
30   Measurement technology, analysis. datacollection, recovery of information
40   Quality and grading criteria, guidelines, objectives
50   Measures for the reduction, prevention or elimination of environmental pollution or their effects
60   Planning aspects
70   Theory, basic principles and general questions, background information

Fig. 2: Subclassification of the Environmental Classification

The environmental classes represent the main topics of the document to be characterized. For example the classes show on one hand the medium which the user is dealing with (water, air, soil etc) and on the other hand the subclassification describes the effects of environmental pollution on this medium.

b) Thesaurus

Our thesaurus is a polyhierarchical one and it is designed for multilingual use. Geographic terms are included in a separate file of the thesaurus, because many problems in the environment are related to regional or local peculiarities. It is structured by broader terms, narrower and related terms. In order to limit the number of controlled terms (descriptors) in our thesaurus we use synonyms and quasi-synonyms. A special way of forming a synonym is the description of a term with a combination of at least two existing descriptors.
Our thesaurus contains about 8100 index terms with additional 22000 nondescriptors (synonyms, quasisynonyms, combinations of terms, single terms (components of terms necessary for automatic indexing), stop words).
The geographic thesaurus and the thesaurus of biological terms comprises more than 3600 descriptors and 2800 nondescriptors.
At the moment our thesaurus and classification are bilingual (german - english).

The environmental thesaurus and classification are used every day by the input teams of our databases and as it were the other side of the databases our clients all over the world use it searching online and retrieving relevant documents.

The environmental thesaurus and classification has not been intended as a comprehensive dictionary, but it should show the terms which are actually used in the different subject fields. Therefore you will find in our thesaurus incomplete hierarchies, which are kept incomplete intentionally. Extensions will be introduced when necessary.

5. Software

Since 1983 we use for managing our databases, the thesauri and the classification a software package called aDIS (i.e. adaptable Documentation and Information System). It was designed as a tool for our mainframe computer we used at that time. Today we are still employing this tool in an updated version and in a client - server - technology. The thesaurus part of aDIS combines traditional approaches of thesaurus development and indexing with the advantages of full text inverted files.

The software allows the analysis of the different texts stored during input activities in our databases. Main aspects of the text analysis function of aDIS are:
- to compare the words of the text with the list of terms in the database
- to elaborate descriptors, composita, combinations of terms in relation to one or more thesauri
- to reduce terms to their root and to code them with the appropriate termination and inflexion
- to control the text with regard to correct orthography
- to divide unknown terms (in particular composita) into controlled single terms of the thesaurus
- to combine identified single terms in a phrase or part thereof to composita of the thesaurus
- to identify synonyms and to substitute these words by descriptors
- to identify the significance of a term by checking the position of the term in the hierarchy and the frequency of occurrence
- to ascertain broader, narrower and related terms if useful
- to find out the correct classification codes for the document.

These functions are applicable for input and retrieval purposes. Regarding the input activities, the use of the above mentioned capabilities of the text analysis program enables us to carry out a semiautomatic indexing (for the german language). 80 % of the terms and classes for indexing suggested by the software are correct and need no intellectual revision.

6. Relations to other activities

a) national

In addition to the demonstrated fields of application at the Federal Environmental Agency, the thesaurus and classification are used for:
- indexing objects in the catalogues of data sources in several german states
- indexing books at many libraries of german environmental agencies
- preparing registers of bibliographies, annual reports, catalogues and publications.

b) international

With Austria there exists an agreement for a joined project, using the thesaurus for the austrian catalogue of data sources and extending it to a multilingual instrument in the german, english, french and italian languages.

The results of this cooperation shall be combined with the activities of the European Environmental Agency to develop a general multilingual environmental thesaurus for all languages in the European Union.

7. Further development

The semiautomatic indexing in connection with our input activities lead to a considerable reduction of necessary manpower per document, thus enabling our staff to prepare a larger number of documents for the databases. We intend to improve this feature of the aDIS software to the english language. First applications in this direction exist in the area of energy. This domain shall be extended to environmental terminology too.

The computer assisted translation of the titles of publications or projects is another aspect to be observed. Due to the capabilities of the aDIS software to operate in two languages and to compare translated documents already stored in a database with new texts will reduce the intellectual effort for individual translations.

The multilingual aspect of environmental thesauri is a very important factor for future considerations. In particular we will have to take into account that the languages of the nations of the eastern part of europe must be included in future projects and activities.

Bruno Felluga, Mario Palmera, Sandra Lucke, Paolo Plini
Reparto Ricerca e Documentazione Ambientale
Istituto Tecnologie Biomediche - CNR, Rome, Italy

A Classification Scheme for a General Multilingual Thesaurus for the Environment

Abstract

In 1991, a project has been started by a working group formed by this Unit, the Centre for Information and Documentation on Environmental Research of TNO (The Netherlands) and Department of the Environment of the United Kingdom, to build a multilingual environmental thesaurus based on the Dutch Milieu-thesaurus, published in 1990. The resulting trilingual (Dutch, English, Indian) "Thesaurus for the Environment" (CNR, 1991) contained about 3000 preferred, post-coordinated and non-preferred terms classified in 30 groups, each group being presented in a hierarchical structure extended to seven levels. A version of this thesaurus, recently enriched with the German equivalents, has been produced on CD-ROM by the Publications Office of the CNR and will be presented during the Conference.

The updated version of the Dutch Milieu-thesaurus (1994), is at present the basic document for the development of a general thesaurus for the environment on the context of an initiative of the European Environmental Agency. The development of a suitable classification scheme for the new thesaurus will include a comparison with the classification schemes of INFOTERRA Thesaurus (1990), of the Umweltbundesamt Umweltklassifikation (1994), of the MOPU Thesauro de medio Ambiente (1990) and of other relevant documents. An analysis is being performed on the classification schemes and on the top terms of the aforementioned documents. It is foreseen that a matrix classification scheme with a thematic and a functional (facetted) axis will be used for the classification of the terms. The expected classification scheme will be in the form of a set of classes, subclasses and top terms defining a lexicon of about 1000 general terms. This classification scheme is going to be used for data entry and information retrieval in databases of environmental data, like the CDS, Catalogue of Data Sources of the European Agency.

Hassane Bendahmane
INFOTERRA PAC, UNEP, Nairobi, Kenya

The INFOTERRA Thesaurus of Environmemtal Terms

Abstract

The paper provides a historical background to the INFOTERRA thesaurus and its evolution; it outlines the objectives of the thesaurus and the constraints dictated by those objectives.

Assuming that "the INFOTERRA thesaurus cannot and should not be everything for everyone", the Author raises the issue of the optimal level of depth (detail) and scope (subject areas) which the INFOTERRA thesaurus should cover. The Author is of the view that the detailed thesauri should be left to the various subject areas with which environment intersects; the scope, however, could be expanded from the environmental construct to the broader sustainable development field, thus providing more room for socio-economic terms.

The author suggests that, while there may be several environmental thesauri to meet special needs of specific user groups, a mechanism should be established to allow for a common outline at the higher levels of aggregation.

65

Christian Galinski
Gerhard Budin, Infoterm, Vienna

Thesaurus and Terminology
Providing Access to Reference Knowledge
in Environmental Information Systems

1. Introduction

In recent months a number of institutions have begun to conceive and implement Catalogues of Data Sources (CDS) for environmental information. The following considerations are directed towards facilitating access to knowledge stored in such environmental information systems by means of indexing languages and terminology databases.

2. New database conceptions: catalogue of data sources (CDS)

The CDS data catalogues contain meta-information providing access to primary information (e.g. individual test data), secondary information (e.g. bibliographical data), or tertiary information (e.g. data on information holdings). These types of information are presented as primary/secondary/tertiary information units (e.g.records) and modelled as information objects having different purposes and/or functions according to the view or needs of the user. By applying different purposes and/or functions, the information objects can become meta-information. In a large-scale meta-information system (such as a CDS in the field of environment) the information objects can be aggregated at different levels of aggregation (e.g. administrative levels) according to pragmatic needs.

It has been soon found out that it is not only useful, but even necessary to include a thesaurus module in the CDS system in order to facilitate access to all kinds of data. The thesaurus module may consist of just one traditional-type thesaurus or of a complex of macro- and micro-thesauri linked together by concordances or other means of linkage. The thesauri used in this context could further be either expanded or linked to reference knowledge in several ways. In order not to allow the individual thesauri to explode beyond manageability by different kinds of reference data, it would be useful to use "natural" terminology (quite different from non-descriptors as they are normally presented in thesauri) as a reference tool.

First of all one has to analyse the types and nature of reference data, which may comprise:

- texts (legal texts, standards, technical rules etc. in the form of full texts with or without pre-indexing)
- individual text occurrences (e.g. clauses of a law, standards etc.)
- classification schemes (e.g. patent classification), data catalogues (e.g. hazardous waste data catalogue) etc.
- names (of institutions and their departments, acitivities, programmes, projects, conventions, etc.).

In some cases (such as a hazardous waste catalogue) these data can itself be the reference data as such and in addition can be used as a micro-thesaurus supplementing the CDS thesaurus module in order to access other reference data (e.g.on companies generating hazardous waste, control institutions, laws and regulations etc.).

In the case of concordances the concordancing can often occur at the same time between individual items of the two thesauri/classification schemes or between section headings, so that the user can easily shift between macro- and micro-levels of information retrieval. It would also be useful to create a hypertext mechanism between the various types of reference data. In the European context this can be achieved by recording and managing thesaurus and terminology data in a very refined and nevertheless open and flexible way in the form of a 'multilingual thesaurus/terminology management' system that is equipped with a multiple link mechanism to reference data.

In certain subject fields, such as environment, extremely complex specifications are required from information systems:

- they shall comprise virtually all kinds of data as completely as possible,
- they should be useful for experts as well as for laypersons,
- politicians at various levels of authority want to use them for decision-making purposes,
- the general public wants to have user-friendly access,
- information is needed in various degrees of aggregation for different purposes, etc.

Users expect information not only

- at the place
- at the time and even
- presented in the form

required, but - an aspect, which so far has been largely neglected - also

- in the language most suitable to them.

The form of information does not only refer just to the user-friendliness, but increasingly also to the degree of abstraction, complexity or detailedness (i.e.aggregation). The language in which information is expected by the user can be either the national language (or a foreign language for comparison purposes) or a different register (or language level, the degree of specialization resp.) with which the user is most familiar.

That is why new approaches have to be conceived, the complexity of which goes far beyond traditional database conceptions. In this contribution environment information is taken as an example for a subject field, in which such new information systems are required at national and multinational (i.e. European) level by many different kinds of user groups.

2. Types of information - an overview

Information can - among others - be subdivided into

- primary information, such as
    texts
    linguistic data (words, phrases etc.)
    terminology
    thesaurus entries
    classes (of a theme classification)
    numerical data
    object-oriented factual data etc.

- secondary information, such as
    bibliographic data (increasingly not only on publications and other
        documents, but also on electronic documents)
    (complex) factual data (e.g. on institutions, programmes etc.)
    tertiary information, especially (referential) factual information such as
    information on holdings
    information on databases or information systems etc.

It becomes obvious that, depending on the context of use, factual data can be primary, secondary or tertiary information.

All kinds of information mentioned above can be further subdivided into many other types. All information can, but not necessarily must be interconnected in some way or other - thus mirroring the cognitive dimension of specialized knowledge. Some data are rather 'volatile', others remain valid over long periods of time. More or less all of them can, and in fact must be used multi-functionally, otherwise the efforts to create this new kind of complex information systems will never pay off (neither from the financial point of view nor from the point of view of user satisfaction). The 'art' of designing such information systems lies in

-   reducing the degree of complicatedness by increasing the degree of
        complexity, while at the same time
-   creating a highly flexible yet simple user interface for the user.

3. Characteristics of the different kinds of information and their representations in specialized discourse

For communication purposes (knowledge transfer) information and knowledge have to be presented in some way, following certain rules and conventions:
Specialized texts are linguistic representations of specialized knowledge. In a database they are treated as documents of various types:

-   legal provisions
-   standards
-   scientific articles
-   (product or process) specifications
-   technical reports
-   political documents
-   theses and dissertations etc.

or parts hereof. Contexts taken from full texts, therefore, also constitute documents in a database.

There are various kinds and types of specialized texts written in different

-   style (different user groups and purposes in mind)
-   languages (for different people)
-   register (for different social groups and different)
-   degree of specificity
-   degree of condensation etc.

Depending on style, register and purpose specialized texts consist of a major or minor share of specialized terminology (60-80% as a rule). Many documents contain terminological documents, such as laws defining the crucial concepts of the respective legal provision, terminology standards (or the clause on "terms and definitions" occurring in nearly all standards), terminological theses and dissertations, etc. Many documents are supplemented by an index providing a quick look-up facility to the user.

Texts can be accessed from a macroscopic perspective by means of documentation languages (which were applied for indexing purposes in order to be able to retrieve the texts) or from a microscopic perspective by means of terminology. Some texts represent specially condensed information, e.g. definitions, definitory contexts, abstracts. While the first two directly refer to units of knowledge (viz. concepts) the latter gives the essence of the text, to which it refers. Increasingly modern specialized texts comprise numerical and graphical or other non-linguistic information.

In the field of environment, numerical data play a very important role. Some of them may be highly volatile, others very long-lasting. Access again is mostly through terminology. Often such data in fact represent (or are/can be combined with otherdata so as to form simple) facts.

With terminology being a structured set of concepts and assigned designations (terms or graphical symbols) of a certain subject field, terminological data may comprise a wide variety of data. Dealing with terminological data for practical purposes is referred to as terminology management and terminology science on the scientific level.

Documentation languages are artificially unified 'controlled vocabularies' applied to documents (in the database) mainly for indexing purposes (indexing language) and retrieval purposes (retrieval language) or knowledge ordering purposes (e.g. classification schemes). Natural language contains too much ambiguity to be efficient for indexing and retrieval. Thesauri, usually representing conceptual structures represented by top classes, top terms and descriptors (supplemented by indications of synonyms or related terms), are the most frequent indexing and retrieval tools.

Bibliographical information comprises data on documents for their identification, retrieval and reference.

Factual data presents data about facts, including all kinds of objects, events, processes. These data may be numerical (primarily statistical), pictorial or textual. MIE (1985, 1991) distinguishes seven types of factual databases:

- referral database
- databases on museum and art objects
- product and method databases
- material databases
- event databases
- databases on regional and demographic data
- statistical databases.

One may add here also geographical information which is important especially in the field of the environment.
Any kind of database containing one of the above mentioned types of data could be extended inprinciple by integrating all other types of data. But usually different creators/users are responsible for different types of information. An all-comprehensive monolithic huge databank would probably soon become unmanageable. It is, therefore, advisable to conceive an integrated, interdisciplinary use of all these information in a modular way (either split up into several databases or combined in one database). The more multi-functional these modules and their data are, the more synergies can be gained.

4. Aggregation

Information aggregates reduce the complexity and diversity of data by abstracting from individual data instances into data types. The degree of aggregation depends on pragmatic requirements, e.g. how much complexity a given information system (database) is able to manage. Aggregated data are mainly found in statistical databases (Staud 1991). The concept of aggregation is not only a basic category of systems theory, but also of computer science (for the latter see Smith/Smith 1977).

In environmental databases, aggregation (together with abstraction, reduction and integration) is a key process of information management. Data models for meta information systems contain iterative aggregation steps. Environmental meta-information systems contain data taken (in a harmonized way) from numerous local databases specializing in some data types. Environmental data catalogues currently implemented (Schütz/Böhm 1994, Schütz/Lessing 1993 for Germany; Keune/Murray/Benking 1991 for HEM and the international level) or planned (on the European level by the EEA, the European Environmental Agency), depend on aggregation techniques in order to reduce the infinite complexity of environmental data on local levels and in order to be able to transform such data collections into data that are of immediate use, e.g. to a minister of the environment. Such meta-data objects may include spatial, temporal, object- and topic-related and mixed aggregates(Schütz/Böhm 1994).

5. Reference knowledge

In principle any kind of data can become reference knowledge in a database or file of another kind of data. In the case of a CDS in the field of environment, where access to information is primarily provided through thesaurus entries and supplemented by a limited amount of terminology (synonyms or quasi-synonyms becoming non-descriptors) the following kinds of information can constitute reference knowledge:

-   terminology (in the form of a more or less systematic glossary, such as a waste-type catalogue, which could be applied for accessing other information related to waste, or in the form of terms and definitions, being knowledge units per definition, and which also could be linked to further related information),
-   text occurrences (from laws, standards, etc. representing a condensed description of one or a few related knowledge units),
-   factual data e.g. on institutions (leading to information on activities, such as programmes and projects etc.), programmes/projects/measures (leading to information on the persons, institutions, groups responsible for them) etc., full texts (of laws, standards, technical rules etc. related to the field of environment).

'Natural' terminology (as it is used in subject-field communication) which normally occurs in thesauri only in the form of non-descriptors together with a USE-pointer to the respective descriptor could be extended in several ways, such as

-   national variants (e.g. Swiss German) or regional variants (e.g. Alpine terminology) with links to other information (e.g. originators, authorites, laws etc.)
-   legal keywords with links to the (secondary) information on the respective law(s) or even to (primary information, i.e.) the legal text(s) itself
-   names of institutions etc.
-   micro-thesauri, such as standardized terminologies, standard nomenclatures (e.g. the chemical terminology of IUPAC, the International Union of Pure and Applied Chemistry), harmonized product catalogues, classification schemes etc., any entry of which could lead to further information stored in different databases, modules or files.

Terms from 'natural' terminologies (incl. nomenclatures etc.) often will correspond to items in a documentation language. But since the function of documentation languages being 'controlled vocabularies' is quite different, it is advisable not to call this correspondence 'equivalency', but rather treat the data realm of terminology differently from that of the documentation languages.

If the term is accompanied by a definition, explanation or other forms of concept description (such as a defining context) the entry as such already constitutes a reference knowledge unit.
Figure 1 contains a definition of environmental data as presented in the Austrian Law on Environmental Information (Schober/Lapatta 1994: 12):

This entry as such can definitely be considered as reference knowledge. A bibliographic reference would lead to the information on the law, in which this entry is contained or even directly to the full text of this law. By means of pointers or other links attached to elements of the definition, the user could search for further clarifications of the contents of the definition.

Umweltdaten

§ 2. Umweltdaten sind auf Datenträgern[1] festgehaltene Informationen[2,3] über

1. den Zustand der Gewässer, der Luft, des Bodens, der Tier- und Pflanzenwelt und der natürlichen Lebensräume sowie seine Veränderungen oder die Lärmbelastung[4];

2. Vorhaben[5] oder Tätigkeiten, die Gefahren für den Menschen hervorrufen oder hervorrufen können oder die Umwelt beeinträchtigen oder beeinträchtigen können[6], insbesondere durch Emissionen, Einbringung oder Freisetzung von Chemikalien[7], Abfällen[7], gefährlichen Organismen oder Energie einschließlich ionisierender Strahlen in die Umwelt oder durch Lärm;

3. umweltbeeinträchtigende Eigenschaften, Mengen und Auswirkungen von Chemikalien, Abfällen, gefährlichen Organismen, freigesetzter Energie einschließlich ionisierender Strahlen oder Lärm;

4. bestehende oder geplante Maßnahmen zur Erhaltung, zum Schutz und zur Verbesserung der Qualität der Gewässer, der Luft, des Bodens, der Tier- und Pflanzenwelt und der natürlichen Lebensräume, zur Verringerung der Lärmbelastung sowie Maßnahmen zur Schadensvorbeugung und zum Ausgleich eingetretener Schäden, insbesondere auch in Form von Verwaltungsakten[8] und Programmen[9].

Figure 1: Entry

This entry as such can definitely be considered as reference knowledge. A bibliographic reference would lead to the information on the law, in which this entry is contained or even directly to the full text of this law. By means of pointers or other links attached to elements of the definition, the user could search for further clarifications of the contents of the definition.

The Austrian Catalogue for Waste (ÖNORM S 2100) is highly systematic in nature, classified by sectors and subsectors, chemical-physical treatment, disposal, thermal treatment, necessary conditioning, eluate classes supplemented by notes. Figure 2 gives an example from this catalogue, explained in the following:

If we have a look at sectors 12 "wastes from vegetable and animal grease products" and 13 "wastes from husbandry and slaughtering" the information on this single page can be used as reference knowledge and reference data in several ways:

- all entries under sector 12 (or 13) may be found also under an equivalent (homonymous or synonymous) entry of the environment thesaurus
- some subsectors, such as 126 "vegetable grease products" could also be found under an equivalent entry of the environment thesaurus
- individual entries, such as 13705 "infectious manure" could be equivalent to an entry of the environment thesaurus
- under a (fictive) thesaurus entry "wastes from vegetable and animal grease products requiring chemical/physical treatment" (A) one could list items 12102, 12302, 12503, 12601, 12702, 12703, 12704 and 12901
- under disposable manure of sub-sector 137 (B) one would only find 13704 with the note "conditioning required"
- under "wastes from vegetable and animal grease products of eluate class IIIb" (C) one would find 12101, 12702, 12703, 12704 and 12901.

This shows how highly structured terminologies/ nomenclatures/ classifications etc. can be used either as individual entries or in subsets of different views in combination with an environment thesaurus. All terminology entries could be linked to bibliographic information (e.g. on legal provisions) or directly to full texts (or text extracts, such as contexts) or names of institutions or organisations concerned with the respective item in some way or other. For data management purposes, however, it is advisable to handle the various types of reference data/ knowledge units in different modules, files or databases.

Schlüssel-Nummer	Bezeichnung	C/P	BB	TB	D	zugeordnete Eluatklasse und Hinweise
12	**Abfälle pflanzlicher und tierischer Fetterzeugnisse**					
121	**Abfälle aus der Produktion pflanzlicher und tierischer Öle**					
12101	Ölsaatenrückstände	–	K	+	K	EK IIIb
12102	verdorbene Pflanzenöle	+	K	+	–	
123	**Abfälle aus der Produktion pflanzlicher und tierischer Fette und Wachse**					
12301	Wachse	–	–	+	–	
12302	Fette (z. B. Frittieröle)	+	K	+	–	
12303	Ziehmittelrückstände	–	–	+	–	
12304	Fettsäurerückstände	–	–	+	–	
125	**Emulsionen und Gemische mit pflanzlichen und tierischen Fettprodukten**					
12501	Inhalt von Fettabscheidern	–	K	K	–	
12502	Molke	–	K	+	–	
12503	Öl-, Fett- und Wachsemulsionen	+	K	K	–	
126	**Produkte aus Pflanzenölen**					
12601	Schmier- und Hydrauliköle, mineralölfrei	+	K	+	–	
127	**Schlämme aus der Produktion pflanzlicher und tierischer Fette**					
12702	Schlamm aus der Speisefettproduktion	+	K	K	K	EK IIIb
12703	Schlamm aus der Speiseölproduktion	+	K	K	K	EK IIIb
12704	Zentrifugenschlamm	+	K	K	K	EK IIIb
129	**Raffinationsrückstände aus der Verarbeitung pflanzlicher und tierischer Fette**					
12901	Bleicherde, ölhaltig	+	–	K	K	EK IIIb
	Ⓐ					
13	**Abfälle aus der Tierhaltung und Schlachtung**					
131	**Schlachtabfälle**					
13101	Borsten und Horn	–	K	K	K	TKV[1])
13102	Knochen	–	K	K	–	TKV
13103	Innereien	–	K	K	–	TKV
13104	Geflügel	–	K	K	–	TKV
13105	Fisch	–	K	K	–	TKV
13106	Blut	–	K	K	–	TKV
13107	Federn	–	K	K	K	TKV, EK IIIb
13108	Magen- und Darminhalte	–	K	K	–	
13109	Wildabfälle	–	K	K	–	TKV
13110	Fleisch- und Hautreste, Därme, sonstige Tierkörperteile	–	K	K	–	TKV
134	**Tierkörper**					
13401	Versuchstiere	–	–	K	–	TKV
13402	Konfiskate	–	–	K	–	TKV
13403	Kadaver	–	–	K	–	TKV
13404	Tierkörperteile	–	–	K	–	TKV
137	**Tierische Fäkalien**					
13701	Geflügelkot	–	K	K	–	
13702	Schweinegülle	–	K	–	–	
13703	Rindergülle	–	K	–	–	
13704	Mist	–	K	K	K	EK IIIb
13705	Mist, infektiös	–	–	K	–	
13706	Kot, infektiös	–	–	K	–	Ⓒ
13707	Gülle, infektiös	–	–	K	Ⓑ	

') siehe Fußnote 1) auf Seite 4

Highly condensed texts, such as definitions, abstracts or defining contexts to which terminology may lead, are reference knowledge units in itself. Their elements (i.e. terminology units contained in these texts) may further point to other reference knowledge units.

Some thesaurus entries and terminology entries may be linked to information on institutions (e.g. regional environment authorities) or organisations (such as waste disposal enterprises), to which other types of information, such as relevance of national environment protection programmes etc., may be linked.

6. Conclusions

A CDS is a sort of a multiple structured hyper-system comprising micro- and macro-structures.

Terminology is linked to thesaurus items by means of multiple concordancing. This isa relatively new dimension in database methodology. As a CDS is a highly complex system, its methodology should be harmonized as much as possible at the national and multinational level from the very beginning of their conception and implementation. Concordances between different thesauri are very difficult and time-consuming to prepare, between such complex information systems as a whole concordancing is next to impossible (or at least not feasible from the economic point of view). Transnational cooperation, therefore, is of utmost importance.

The problems created by multilingualism can also be solved best by joint efforts. A meta-harmonization tool for database management is currently being created by ISO, UN/ECE and other organizations at the international level: the Basic Semantic Repository (BSR). It is based on an harmonized treatment of data by applying uniform methods of data element definitions, creation of data dictionaries and cross-application data management. Further standardization needs will arise in the course of the further development of CDS.

The combination of methods of terminology work with those of documentation - recently called "T&D" (Galinski 1991, Galinski/Budin 1993) provides the methodology for an efficient preparation, adaptation and use of multilingual thesaurus and terminologies. The difference between thesauri as representing the macro-structure of knowledge and the terminology as representing the micro-structure of knowledge needs not (or even should not) to be made explicit to the user (at least not, if it is not needed). But in the system a clear distinction has to be made between the two kinds of knowledge structures. Via the thesaurus and terminology entries at various levels of abstraction the user can have user-friendly access to environment information by using his/her language.

It need not be stressed that information systems, such as CDS need a clearcut design and methodology. Environmental Information Management must include the management of complexity of specialized information and - on the cognitive level - of specialized knowledge by adopting a modular, flexible and multi-functional approach to knowledge modelling by integrating knowledge organization and terminology management methods (Budin 1993). Methodology, therefore, must have priority over the concentration on tools.

7. References

Budin 1993: Budin, Gerhard. Knowledge, Organization and Modelling ofTerminological Knowledge. In: Schmitz. Klaus-Dirk [ed]. TKE 93: Terminology and Knowledge Engineering. Proceedings Third International Congress on Terminology and Knowledge Engineering. Cologne, 25-27 August 1993, Cologne, Germany. Frankfurt a.M.: INDEKS 1993, p. 1-7.

Galinski 1991: Galinski, Christian. From "terminology documentation" (TD) to "terminology &

documentation" (T&D) - T&D as a prerequisite of information management. TermNet News 32/33 (1991), p. 7-14

Galinski/Budin 1993: Galinski, Christian; Budin, Gerhard. Comprehensive Quality Control in Standards Text Production and Retrieval. In: Strehlow, Richard; Wright, Sue Ellen [eds.]. Standardizing Terminology for Better Communication: Practice, Applied Theory, and Results. Philadelphia: ASTM, 1993, p. 65-74

Keune/Murray/Benking 1991: Keune, Hartmut; Murray, Beatrice, A.; Benking, Heiner. Harmonization of Environmental Measurement. GeoJournal 23.3 (1991), p.249 255

Mie 1985: Mie, Friedrich: Zur Terminologie und Typologie von Fakteninformationssystemen. Nachrichten für Dokumentation 36 (1985)No. 2, p. 66-72

Mie 1991: Mie, Friedrich. Fakteninformationsysteme. In: Buder, Marianne; Rehfeld, Werner; Seeger, Thomas [eds.]. Grundlagen der praktischen Information und Dokumentation. 3. neu gefaßte Auflage. München/London/New York/Paris: Saur,1991, p. 547-558

ÖNORM S 2100 Abfallkatalog, 1990

Schober/Lopatta 1994: Schober, Walter; Lopatta, Hans [eds.]. Umweltinformationsgesetz mit Anmerkungen. Wien: Verlag Österreich/Österreichische Staatsdruckerei, 1994

Schütz/Böhm 1994: Schütz, Thomas; Böhm, Regine. Die Datenstrukturierung des Metainformationssystems Umweltdatenkatalog. In: Kremers, H. [ed.]. Umweltdatenbanken. Marburg: Metropolis-Verlag, 1994

Schütz/Lessing 1993: Schütz, Thomas; Lessing, Helmut. Zum Datenmodell des Umwelt-Datenkataloges Niedersachsens. Metainformation von Umwelt-Datenobjekten. In: Jaeschke, A.; Kömpke, T.; Page, B.; Radermacher, F. [eds.]. Informatik für den Umweltschutz. 7. Symposium Ulm. 1993

Smith/Smith 1977: Smith, J.M.; Smith, D.C.P. Database Abstractions: Aggregation and Generalization. In: ACM Transactions on Database Systems, 1977, vo. 2, no. 2, p. 105-113

Staud 1991: Staud, Josef L. Statistische Information. In: Buder, Marianne; Rehfeld,Werner; Seeger, Thomas [eds.]. Grundlagen der praktischen Information und Dokumentation. 3. neu gefaßte Auflage. München/London/New York/Paris: Saur,1991, p. 402-427

Pavla Stančikova, Jan Gotthard, Marek Smihla
CEIT, Centre of ECO-Information and Terminology
Bratislava, Slovakia

Linking Databases on Standards and Terminology - A Basis for Electronic Publication Processing on Environmental Topics

Abstract: The Micro CDS/ISIS software package was tested from the point of view of providing linkage functions of various databases for integrated search purposes. The database BATON was designed under the software CDS/ISIS and filled with converted data on standards available at Vienna's Austrian Standards Institute WANG system (9830 records). Simultaneously the terminology database TATON was designed under Micro CDS/ISIS software and filled with converted terminology data available at the Austrian Standards Institute (INFOTERM) WANG system (30310 records). The linkage of both databases BATON and TATON was programmed to the standard Micro CDS/ISIS in order to extend search possibilities of both databases and to allow linked searches. This function had not been available on the WANG system.

The linked BATON and TATON databases are available at INFOTERM Vienna. They are not only used for regular queries in both databases but also for preparation of publications in various subject fields. The subject fields of 'Civil Engineering' and the 'Construction' were tested as well as environmental topics such as 'Urban Hygiene', 'Wastes', 'Public Health Engineering', 'Water', etc. This contribution gives some examples of results form the environmental field based on the tool of linked databases and standards and terminology as extension of the standard design of Micro CDS/ISIS software.

1. Background

The databases BATON and TATON were designed under the software Micro CDS/ISIS. The access to the databases is menu driven (See Fig.1). An advanced programming file allowed to make linkages between the databases BATON and TATON. The main menu of BATON and TATON application has standard functions with a slightly changed layout (See Fig.2). Linkages between BATON and TATON were incorporated into the menu on the second level with additional keys C for a linked search in the TATON database and viceversa and M for moving from one database to the other (See Fig.3).

2. The Database BATON

The BATON database (former DON) is the bibliographic database of the ÖNORM standards. The record structure was designed in the same way as the one on the WANG system at the Austrian Standards Institute. From the retrieval point of view the following fields have a key function:

- key for linkage with the terminology database TATON
- document number, for example ÖNORM A 1610 T 2
- UDC notation
- descriptors

3. The Database TATON

The TATON database (former TON) is the ÖNORM terminology database. The record structure was designed to be identical with the one operated on the WANG system at the Austrian Standards Institute. In the TATON database the following fields can be searched:

- key for linkage with BATON database
- document number
- entry term DE (German)
- entry term EN (English)
- entry term FR (French)
- entry term RU (Russian)

4. Linked BATON and TATON Databases

CEIT Bratislava - as the designer of both databases as well as of the linked version after the study of the record structures of both databases - had to take into accout the following:

- key to key linkage means record to record linkage
- identical fields are only keyed for linkage and document number
- the database BATON focusses on standards and their document numbers
- the database TATON focusses on terms which may appear in different standards
- a UDC notation as well as descriptors can be used to form subject sets of ÖNORM standards
- subject sets of terms can be created only as a linked search from the BATON database (as there is no classification system in the TATON database)
- further processing of a subject set created in the BATON database gives more opportunities for electronic publishing and publication processing of terminology sources from different subject fields.

5. Methodology of Linked Searches in the BATON/TATON Databases

A decision was made to demonstrate at the conference on *Environmental Knowledge Organization and Information Management* the linked search in the subject field 'Urban hygiene. Wastes' with the notation 628.4. The Micro CDS/ISIS standard retrieval functions allow to:

- select the desired subject from the vocabulary
- use the precise search formulation function
- combine a search formulation with a right truncation

In this special case, the combination of the search formulation with a right truncation was the most suitable method.

For our purposes we use the following search formulation:

628.4$

The BATON database has 48 records on the topic of 'urban hygiene' and 'wastes'. A sample of the search result in the standard BATON display format is given in Fig.4.

The search in the BATON databases was followed by the linked search in the TATON ones. The search result of the linked search on terms in the subject field of 'urban hygiene' and 'wastes' resulted in 281 records. The example of the result of a linked search in the TATON database in the standard TATON display format is given in Fig.5.

Micro CDS/ISIS is able to process indexes. A sample of processed terms of the search result in the TATON database is given in Fig.6. For any publication this is a very powerful feature of the BATON/TATON linked databases: the ability to provide an index of terms with the indication of the document number of a standard.

For reasons of comparison another search on the topics of 'urban hygiene' and 'wastes' was processed in the BATON database. The index of used descriptors thus created for the description of the contents of standards can be seen in Fig.7.

6. Conclusion

We wanted to point out in this contribution that the Micro CDS/ISIS software is a very flexible and open system, able to extend the standard design of databases to a more powerful one and by this to create a new generation of tools available for electronic publishing of terminology information sources. The examples of publications presented show also the results of the interface of Micro CDS/ISIS with the DesktopPublishing system VENTURA. The system described is installed at INFOTERM Vienna as well as at CEIT Bratislava and is ready for multiple scientific applications.

7. References

(1) Stančikova, P., Gotthard, J., Smihla, M.: Micro CDS/ISIS Introduction Manual. Bratislava: CEIT 1994. 34p.

(2) Stančikova, P., Gotthard, J., Smihla, M.: Micro CDS/ISIS Application Manual for Information Sources in Terminology. Databases BATON and TATON. Bratislava: CEIT 1994. 29p.

Fig.1: Menu driven access to databases

Fig.2: The main menu of BATON and TATON

Fig.3: Linkages between BATON and TATON

Fig.4: Sample of search result in standard BATON display format

Fig.5: Sample of the result of linked search in the TATON database

Fig.6: Sample of processed terms of the search result in the TATON database

Fig.7: Created index of descriptors used for description of standard content.

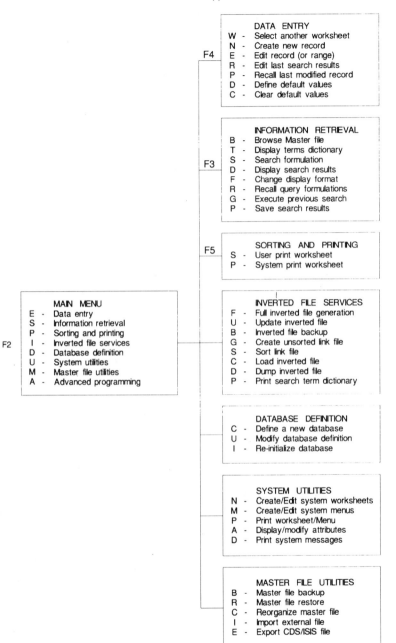

Fig. 1: Menu driven access to databases.

```
**********************************************************
********** Micro CDS/ISIS  -  Version 3.0 **********
**********************************************************

        ┌──────────────────────────────────────────────┐
        │ APPLICATION ON STANDARDS AND TERMINOLOGY     │
        │             DATABASES                        │
        └──────────────────────────────────────────────┘

                  C - Change data base
                  L - Change dialogue language

        ┌──────────────────────────────────────────────┐
        │ E - ISISENT - Data entry services            │
        │ S - ISISRET - Information retrieval services  │
        │ P - ISISPRT - Sorting and printing services   │
        │ D - ISISDEF - Data base definition services   │
        │ M - ISISXCH - Master file services            │
        │ U - ISISUTL - System utility services         │
        │ A - ISISPAS - Advanced programming services   │
        └──────────────────────────────────────────────┘

                  X - Exit (to MS-DOS)
                           ?

Data base: BATON                              Worksheet: BATON
Max MFN  : 102                                Format  : BATON
            Micro CDS/ISIS - (C)Copyright Unesco 1992
```

```
**********************************************************
********** Micro CDS/ISIS  -  Version 3.0 **********
**********************************************************

        ┌──────────────────────────────────────────────┐
        │ APPLICATION ON STANDARDS AND TERMINOLOGY     │
        │             DATABASES                        │
        └──────────────────────────────────────────────┘

                  C - Change data base
                  L - Change dialogue language

        ┌──────────────────────────────────────────────┐
        │ E - ISISENT - Data entry services            │
        │ S - ISISRET - Information retrieval services  │
        │ P - ISISPRT - Sorting and printing services   │
        │ D - ISISDEF - Data base definition services   │
        │ M - ISISXCH - Master file services            │
        │ U - ISISUTL - System utility services         │
        │ A - ISISPAS - Advanced programming services   │
        └──────────────────────────────────────────────┘

                  X - Exit (to MS-DOS)
                           ?

Data base: TATON                              Worksheet: TATON
Max MFN  : 75                                 Format  : TATON
            Micro CDS/ISIS - (C)Copyright Unesco 1992
```

Fig. 2: The main menu of BATON and TATON.

```
Service ISISRET          Information Retrieval Services   Menu EXGEN


                 ┌─────────────────────────────────────┐
                 │ L - Change dialog language          │
                 │ B - Browse Master file              │
                 │ T - Display terms dictionary        │
                 │ S - Search formulation              │
                 │ D - Display search results          │
                 │ G - Execute previous search         │
                 │ F - Change display format           │
                 │ R - Recall query formulations       │
                 │ P - Save search results             │
                 │                                     │
                 │ C - Linked search in TATON datab.   │
                 │ M - Move to TATON database          │
                 └─────────────────────────────────────┘
                   X - Exit

                             ?   _

Data base: BATON                            Worksheet: BATON
Max MFN  : 102                              Format : BATON
                Micro CDS/ISIS - (C)Copyright Unesco 1992
```

```
Service ISISRET          Information Retrieval Services   Menu EXGEN


                 ┌─────────────────────────────────────┐
                 │ L - Change dialog language          │
                 │ B - Browse Master file              │
                 │ T - Display terms dictionary        │
                 │ S - Search formulation              │
                 │ D - Display search results          │
                 │ G - Execute previous search         │
                 │ F - Change display format           │
                 │ R - Recall query formulations       │
                 │ P - Save search results             │
                 │                                     │
                 │ C - Linked search in BATON database │
                 │ M - Move to BATON database          │
                 └─────────────────────────────────────┘
                   X - Exit

                             ?   _

Data base: TATON                            Worksheet: TATON
Max MFN  : 75                               Format : TATON
                Micro CDS/ISIS - (C)Copyright Unesco 1992
```

Fig. 3: Linkages between BATON and TATON.

```
**** 693 ****     00017603
DOKNR =ONORM M  9463
DOKART=N
AUSG  =1992 09 01
EMPDAT=
DTITEL=Verbrennungsanlagen fuer Abfall aus dem medizinischen Bereich,
Durchsatzleistung bis 350 kg/h - Technische Anforderungen,
Emissionsbegrenzungen
ETITEL=Incinerators for waste from medical institutions, capacity up
to 350 kg/h - Technical requirements, emission limits
FTITEL=Installations d'incineration des dechets du domaine medical,
capacite jusqu'a 350 kg/h - Exigences techniques, limites d'emission
EINSPR=
UDC   =628.474.373:61.002.68
HRSG  =Osterreichisches Normungsinstitut  Heinestrasse 38   A-1020
Wien
ZITAT =OENORM B  3800 T  2 * OENORM C  1108 * OENORM C  1109 * OENORM
       M  5861 * OENORM M  7515 * OENORM M  7531 * OENORM M  7532 *
       OENORM M  9415 T  1 * OENORM M  9440 * OENORM S  2100 * OENORM
       S  2104 * OEAL-Richtlinie    28 * VDI  2459 Blatt 1 * VDI
       2459 Blatt 4 * VDI  2459 Blatt 6 * VDI  2462 Blatt 1 * VDI
       2462 Blatt 2 * VDI  2462 Blatt 3 * VDI  2462 Blatt 4 * VDI
       2462 Blatt 5 * VDI  2462 Blatt 6 * VDI  2470 Blatt 1 * VDI
       3480 Blatt 1 * VDI  3481 Blatt 1 * VDI  3481 Blatt 2
ERSATZ=OENORM M  9463:1991 05 01OENORM M  9463:1982 09 01
IDENT =
AUCHIN=99 * M * K * U2 * 51 * S * S3
DESK  =Verbrennungsanlage; Sonderabfall; Medizin; Luftreinhaltung;
       Abfallbeseitigung; Abfallwirtschaft; Abfall; Anforderung;
       Emissionsbegrenzung; Begriffe; Terminologie;
       Krankenhausabfall; Lagerung; Transport; Abfallsammlung;
       Verbrennung; Messung; Messbedingung; Emissionswert;
       Messverfahren; Ueberpruefung; Betriebsueberwachung;
       Umweltschutz; Messbericht; Sicherheitsanforderung
REGIST=Verbrennungsanlage; Sonderabfall; Medizin; Abfall
SACHGR=4850
AUTOR =FNA 139 Luftreinhaltung
PREISG= 10
REF   =Diese OENORM enthaelt Anforderungen an Anlagen zur Verbrennung
von Abfaellen der Human\und Veterinaermedizin, die in Kliniken,
Krankenhaeusern, Sanatorien, medizinischen Instituten
(einschliesslich Forschungsinstituten), Pflegeheimen, Arztpraxen,
Ambulatorien und sonstigen Einrichtungen des Gesundheitswesens
anfallen. Diese OENORM ist nicht auf radioaktive Abfaelle anzuwenden.
Es werden sowohl Verbrennungsanlagen als auch Pyrolyseanlagen
beschrieben.
GSG   =
UEBERS=D
SEITEN=007
PARA  =


**** 759 ****     00017857
DOKNR =ONORM S  2000
DOKART=N
AUSG  =1992 10 01
EMPDAT=
DTITEL=Abfall - Benennungen und Definitionen
ETITEL=Waste - Terms and definitions
FTITEL=Dechets - Termes et definitions
EINSPR=
UDC   =628.4.04.001.11
HRSG  =Osterreichisches Normungsinstitut  Heinestrasse 38   A-1020
Wien
ZITAT =
ERSATZ=OENORM S  2000:1991 11 01OENORM S  2000:1986 01 01
IDENT =
AUCHIN=99 * S * S3 * U2 * 51
DESK  =Abfall; Benennung; Definition; Abfallwirtschaft; Begriffe;
```

```
         Muell; Terminologie
REGIST=Abfall; Benennung; Definition
SACHGR=4850
AUTOR =FNA 157 Abfallwirtschaft
PREISG=   3
REF   =In dieser OENORM werden unter Beruecksichtigung der
Gesetzgebung in Oesterreich wesentliche, in der Abfallwirtschaft
gebraeuchliche Begriffe definiert.
GSG   =
UEBERS=D
SEITEN=002
PARA  =
```

Fig. 4: Sample of search result in standard BATON display format.

```
                    WASTES MANAGEMENT

                    Database TATON

     database design and processed by : CEIT Bratislava, Slovakia

00110                   00015649
BENEND=Abfall
BENDEF=001*D *Stoffe, deren sich der Besitzer entledigen will oder
             entledigt hat oder deren Behandlung durch besondere
             Vorschriften geregelt ist.
BENSTW=
DOKNR =ÖNORM S  2100

00112                   00029772
BENEND=Abfall
BENDEF=001*D *bewegliche Sachen, deren sich der Eigentümer oder
             Inhaber entledigen will oder entledigt hat oder deren
             Erfassung und Behandlung als Abfall durch
             Rechtsvorschriften geregelt ist.
       002*A *Der Abfallbegriff schließt neben festen Stoffen
             grundsatzlich auch flüssige und gasformige Abfalle ein.
BENSTW=
DOKNR =ÖNORM S  2000

00113                   00029878
BENEND=Abfall
BENDEF=001*D *bewegliche Sachen, deren sich der Eigentümer oder
             Inhaber entledigen will oder entledigt hat oder deren
             Erfassung und Behandlung als Abfall durch
             Rechtsvorschriften geregelt ist.
BENSTW=
DOKNR =ÖNORM S  2104

00115                   00029626
BENEND=Abfall aus dem medizinischen Bereich
BENDEF=001*U *Abfall aus Einrichtungen, die dem AIDS-Gesetz,
             Apothekengesetz, Dentistengesetz, Hebammengesetz,
             Krankenanstaltengesetz, Plasmapheresegesetz oder
             Tierarztegesetz unterliegen, sowie aus medizinischen
             und veterinarmedizinischen Versuchs-, Untersuchungs-
             und Forschungsanstalten.
BENSTW=BE:medizinischer Bereich
DOKNR =ÖNORM M  9463
```

Fig. 5: Sample of the result of linked search in the TATON database.

WASTES MANAGEMENT

Database TATON - Index of Terms

Designed and processed by: CEIT Bratislava

Abfall

ONORM S 2100, ONORM S 2000, ONORM S 2104

Abfall aus dem medizinischen Bereich

ONORM M 9463, ONORM S 2000, ONORM S 2104

Abfall-Zwischenlagerung

ONORM S 2001

Abfallager auf Zeit

ONORM S 2005

Abfallbereitstellung

ONORM S 2001, ONORM S 2104

Abfallcharge

ONORM S 2111

Abfalldesinfektion

ONORM S 2104

Abfallendbehandlung

ONORM S 2003, ONORM S 2001

Abfallsammlung

ONORM S 2001

Abfallstoffe

ONORM S 2006

Abfalltransport

ONORM S 2001

Abfallvermeidung

ONORM S 2001

Abfallverwertung

ONORM S 2001, ONORM S 2006

Abfallwirtschaft

ONORM S 2001

Ablagerungsdichte

ONORM S 2005

Abrollstrecke

ONORM S 2005

Fig. 6: Sample of processed terms of the search result in TATON database.

WASTES MANEGEMENT

Database BATON

Designed and processed by: CEIT Bratislava

Abbildung

```
ONORM S  2015, ONORM S  2018, ONORM S  2019, ONORM S  2044,
ONORM S  2065, ONORM S  2066, ONORM S  2073,
ONORM S  2074 T  2, ONORM S  2076
```

Abdichtung

```
ONORM S  2074 T  2
```

Abfall

```
ONORM M  9463, ONORM S  2000, ONORM S  2001, ONORM S  2002,
ONORM S  2005, ONORM S  2006, ONORM S  2018, ONORM S  2019,
ONORM S  2023, ONORM S  2052, ONORM S  2066, ONORM S  2070,
ONORM S  2071, ONORM S  2072, ONORM S  2075, ONORM S  2100,
ONORM S  2101, ONORM S  2104, ONORM S  2111, ONORM S  2200,
ONORM S  2201
```

Abfallablagerung

```
ONORM S  2003, ONORM S  2110
```

Abfallbehandlung

```
ONORM S  2003, ONORM S  2004, ONORM S  2005, ONORM S  2006,
ONORM S  2018, ONORM S  2019, ONORM S  2065, ONORM S  2103,
ONORM S  2104
```

Abfallbeseitigung

```
ONORM M  9463, ONORM S  2001, ONORM S  2002, ONORM S  2003,
ONORM S  2004, ONORM S  2005, ONORM S  2006, ONORM S  2013,
ONORM S  2014, ONORM S  2015, ONORM S  2016, ONORM S  2018,
ONORM S  2019, ONORM S  2031, ONORM S  2032, ONORM S  2033,
ONORM S  2037, ONORM S  2040, ONORM S  2051, ONORM S  2065,
ONORM S  2066, ONORM S  2071, ONORM S  2072, ONORM S  2073,
ONORM S  2074 T  1, ONORM S  2074 T  2, ONORM S  2076,
ONORM S  2101, ONORM S  2103, ONORM S  2104, ONORM S  2200
```

Abfallcontainer

```
ONORM S  2015, ONORM S  2065, ONORM S  2066
```

Abfallendbehandlung

```
ONORM S  2003
```

Abfallkatalog

```
ONORM S  2100
```

Abfallsammlung

```
ONORM M  9463, ONORM S  2002, ONORM S  2013, ONORM S  2014,
ONORM S  2015, ONORM S  2016, ONORM S  2065, ONORM S  2066,
ONORM S  2103
```

Abfallverwertung

```
ONORM S  2006
```

Fig. 7: Created index of descriptors used for description of standard content.

Ingetraut Dahlberg
International Society for Knowledge Organization (ISKO), Frankfurt/Main

Environment-Related Conceptual Systematization

Abstract: The environmental sciences can be characterized as being of a multidisciplinary nature and therefore their concepts must be looked at as only partly belonging to its own area. The majority of its concepts are to be "borrowed" from other fields of knowledge and integrated into a consistent framework. These "other" fields are all those having a relationship to the characteristics embedded in the concept of "environment".
The theoretical foundations of concepts and concept systems are roughly outlined. As a practical approach to environmental concept systematization a proposal is made for a conceptual system that takes care of the circumstances mentioned. Its structure comprises the elements of the genuine area of the environment and offers system positions for the interconnection of all the other fields into this structure. This structure is open and heuristsic, it allows for any kind of concept combinations occurring and necessary in the environmental sciences.

1. Why do we need to care about order?

Jesus Christ once told his disciples the following story[1]: There was once a poor farmer. On his acres there were many stones, he could hardly manage to grow something on them. So this farmer decided to collect these stones and put them altogether on 10 different heaps according to their different sizes. His neighbours watched him doing this and loughed about this - in their minds - foolish behaviour. One day, however, an architect came along, saw the fine order of stones and - as he needed just those stones in their good order - asked the farmer whether he may have them and offered a fine price. The farmer happily got rid of his stones while his neighbours' faces became long and green of anger. Thus, *to create order will bear fruits* - if not immediately, then eventually.

In another sense we humans depend on order, viz. our brain functions like a system, like a network of items and establishes immediately a relationship between a known item and a new one for the understanding and handling of any new item, thus it uses its network and its ability to relate and to extend its store of known items. *Our brains establish order for understanding.*

We are indeed not only planned and created to function consciously and unconsciously properly and orderly, we also *enjoy the order in our lives:* the order of night and day, rest and work, the order of hunger and eating, of longing and fulfilment, etc. And also the *happiness in creating order* of any kind.

In our childhood already we learned that certain things must be put at certain places and that - if removed - they must be returned to their places. Mother told us already: *"Order is a necessity for any retrieval purpose!"*

Quite the opposite is propagated today in the computer dependent world: People started to maintain, these mighty instruments are capable to find anything in a free text for which one would be looking. No prethinking would be necessary about the concepts to be searched for, they can be found via one's own language capabilities, "just search for the terms that come to your mind!" What an illusion! R.Fugmann, who already 32 years ago published an article entitled "Order - supreme command in documentation" (1) reminds us in many articles and in his recent books (2, 3) that very soon frustration will be felt if one would depend on words or terms for the identification of concepts in a text. Concepts which alone contain the knowledge one is looking for can be expressed in many different forms and even do not need to be stated expressedly!

In our times technology has taken possession of the minds of people and mankind has been able to make things so very much easier to handle than ever before. We do not calculate anymore in our heads, we have our little pocket calculators. We can - of course - make use of facilities invented but we should not forget to use our brains for what they have been given to us and therefore let us now strain them a bit to look for solution towards the problems of Environment-related Conceptual Systematization!

2. What are Concepts and Concept Systems?

The question about concepts has been with mankind since at least 2500 years and has been answered by many in many ways. In our frame of reference we look at *concepts as units of knowledge*. In earlier publications I used the model of a triangle, stating that for a concept to come into being one needs to have something to refer to on the one hand and true and essential statements about this referent on the other hand. The synthesis of these statements (predications) into one name, term, code fixes these for communicative purposes. Each of these statements - which are judgements/propositions results in a characteristic of the referent in question, we call it also *a knowledge element*. And the sum of all knowledge elements attributed to a referent will then make up *a knowledge unit*, generally spoken of as *the concept.*

Knowledge elements/characteristics are always only components, parts of concepts, e.g., to state "something is colourful" depends on that something which can have a colour. 'colour' in itself can be referred to as a 'property' expressed by the term 'colour', 'Farbe', 'couleur', etc. and as such is a concept by itself just as all the kinds of coulours man has found to exist and to name. Similarly we can state that a "chair is a kind of furniture". Here, 'kind of furniture' is one of the characteristics to be stated of a chair, but 'furniture' by itself is a concept of its own. From these examples it can already be derived that there are different kinds of characteristics which can possibly be stated of a referent. Indeed the sum of all true and necessary characteristics to be stated of a referent make up a little system of its own: the material contents of a concept. To show this difference between a concept looked at (1) formally and (2) materially we would like to use the following triangle and circle:

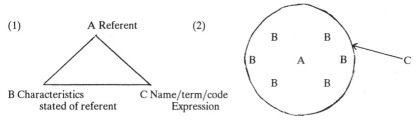

(1)   A Referent   (2)

B Characteristics   C Name/term/code
   stated of referent   Expression

Fig.1: Concept Triangle and Concept Circle to show (1) the formal and (2) the material approach to concepts.

Why is it necessary to recognize a knowledge unit as a set or better as a little system of characteristics? The answer is very easy: Because in this way we can understand that the common possession of the same characteristics in different concepts/knowledge units holds for the relationshipss existing between different concepts. And these relationships are indeed responsible for the construction of different concept systems. In earlier publications (4, 5) I distinguished four kinds of concepts systems: the generic, partitive, opposition/complementary, and functional ones (also included in the German Standard DIN 32 705) (6). It would need too much space to explain these here again in detail. But what is extremely necessary to know about the characteristics of concepts is their relationship to categories. We saw above in the examples of colour and chair that two kinds of categories are implied here: properties and objects. These, as well as the categories of activities and dimensions belong to the socalled 'form categories' already found by Aristotle. But he distinguished also socalled 'levels of being' and subsequently their categories, viz.:

> dead matter (stones, celestial bodies, etc.)
> living beings (plants, animals)
> spiritual beings (humans)
> divine being (God).

In giving the examples in brackets we already combined these levels of being with the form category of 'objects' and by this combination we created - what I called once (7) in opposition to categories of form and categories of being - subject categories (Sachkategorien).

3. An Ontologically-based General Concept System

In extending Aristotle's first 3 levels of being somewhat we arrive at 9 ontical levels which we could use - and have done so - as the basis for a new universal classification system of knowledge fields named the Information Coding Classification, ICC (8). It is shown in Fig.2.

It can easily be seen that the main column on the left constitutes an evolutionary sequence of 3 x 3 main object areas. The first 3 areas of 'Form and Structure', 'Matter and Energy', and 'Cosmos and Earth' would comply with what Aristotle called "Dead matter", although we are not sure any longer whether matter is so 'dead' as not to have internally extremely active movements, etc.

The second 3 areas, the Bio-Area, the Human Area and the Socio-Area would relate to the next two of Aristotle's levels whereas the last 3 areas have not been considered by him as he proceeded only from what was there but not from what man and society made out of nature and his and its capabilities, viz. material, intellectual and spiritual products here in the third 3 areas, Areas 7, 8 and 9, which have just as much an ontological value as the pure objects and their products found as provided by nature of which man is a part too, however gifted with the divine spirit and able to create according to his own free will. (Therefore, the quality of what man creates and produces depends very much on the quality of the use of his freedom.)

The 9 ontical areas thus understood can be looked at under 9 aspects which we called the 'systematifier' principle (10). It may be equalled to what Tony Judge called his 'Structural Outliner' (9). It is a set of 3 x 3 facets relating to form categories which can be explained in the following way:

1   General and theoretical concepts of a knowledge field
2   Object-related concepts (objects, kinds of objects, their elements, parts, etc.)
3   Activity-related concepts (states, processes, operations)
4-6 Concepts related to specialties of objects and/or activities contained in facets 2 and 3
7   Concepts of influences on the previous facets from outside
    ("instrumental", technical relationship)
8   Concepts of the use of previous facets in other fields
    ("potential", resource orientation, maintenance, application)
9   Concepts concerning knowledge about the previous facets when distributed by persons,
    societies, documents, or used for educational purposes and in other kinds of applications
    ("actualization", synthesizing, 'environmental' relationship)

Fig.3: The 'Systematifier' of nine facets to subdivide any knowledge area or field

It has been observed that the application of this Systematifier to the ICC on the level of general object areas cannot be performed as strictly as on the next level, viz. of the subject groups. Therefore, the categorial concepts on top of the nine general object areas do not apply correctly in each case of columns created by them. Whether this is a failure of the system constructor or of the categories or of the levels must be judged by

0 GENERAL FORM CONCEPTS	01 THEORIES, PRINCIPLES	02 OBJECT, COMPONENT	03 ACTIVITY, PROCESS	04 PROPERTY ATTRIBUTE	05 PERSONS OR CONT'D	06 INSTITUTION OR CONT'D	07 TECHNOLOGY & PRODUCTION	08 APPLICATION & DETERMINAT.	09 DISTRIBUTION & SYNTHESIS
1 FORM & STRUCTURE AREA	11 Logic	12 Mathematics	13 Statistics	14 Systemology	15 Organization Science	16 Metrology	17 Cybernetics, Control & Automation	18 Standardization	19 Testing and Monitoring
2 ENERGY & MATTER AREA	21 Mechanics	22 Physcis of Matter	23 General and Technical Physics	24 Electronics	25 Physical Chemistry	26 Pure Chemistry	27 Chemical Technology& Engineering	28 Energy Science and Technology	29 Electrical Engineering
3 COSMO & GEO-AREA	31 Astronomy & Astrophysics	32 Astronautics & Space Research	33 Basic Geosciences	34 Atmospheric Sciences & Technology	35 Hydrospheric &Oceanol.Sci &Technology	36 Geological Sciences	37 Mining	38 Materials Science & Technology	39 Geography
4 BIO-AREA	41 Basic biological Sciences	42 Microbiology and Cultivation	43 Plant Biology and Cultivation	44 Animal Biology and Breeding	45 Veterinary Sciences	46 Agriculture & Horiculture	47 Forestry & Wood Sci. & Technology	48 Food Science and Technology	49 Ecology and Environment
5 HUMAN AREA	51 Human Biology	52 Health and Theoretical Medicine	53 Pathology and Practical Medicine	54 Clinical Medicine & Cure	55 Psychology	56 Education	57 Profession Sci., Labor, Leisure	58 Sport Science and Sports	59 Household and Home Life
6 SOCIO AREA	61 Sociology	62 State and Politics	63 Public Administration	64 Money and Finances	65 Social Aid, Social Politics	66 Law	67 Area Planning, Urbanism	68 Military Science and Technology	69 History Science and History
7 ECONOMICS & TECHNOLOGY AREA	71 General and National Economics	72 Business Economics	73 Technology in general	74 Mechanical & Precision Engineering	75 Building	76 Commodity Science & Technology	77 Vehicle Science and Technology	78 Transportation Technology & Services	79 Utilities and Service Economics
8 SCIENCE & INFORMATION AREA	81 Science of Science	82 Information Science	83 Informatics, computer sience	84 Information in general	85 Communicat. Science	86 Mass Communication	87 Printing and Publishing	88 Communicati.n Engineering	89 Semiotics
9 CULTURE AREA	91 Language and Linguistics	92 Literature and Philology	93 Music and Musicology	94 Fine Arts	95 Performing Arts	96 Culture Sciences, narrower sense	97 Philosophy	98 Religion and Secret Teachings	99 Christian Religion

Fig. 2 · Information Coding Classification. Survey of Subject Groups.

© 1982, rev. 1992 I.Dahlberg

the user. However, in 80% of the cases they can be applied whereas only in one of the 81 subject groups the systematifier could not be applied, viz. in 92 Literature and Philology.

4. The Systematifier in Action

Let us now turn to that subject group in our ICC which relates to our conference: *49 Ecology and Environment*, placed at the synthesizing position 9 at the end of Bio Area 4.

First of all we need to analyze and define (as always when applying the Systematifier to any area or field of knowledge) what is meant by these two concepts and in what sense they are related to each other to form just one subject group.

> *Ecology (de: Ökologie/Umweltbiologie) is the science of the relationship of organisms among each other and to their environment. It comprises biogeography, biosociology, geobiology, pedobiology, hydrobiology, landscape biology, forest biology etc. and includes also the biological foundations of environmental protection, environmental hygiene a.o.*

> *Environmental Science (de: Umweltwissenschaft, Humanökologie, Environtologie, Zivilisationsökologie) is the science of the relationship of the human organism to natural factors (air, temperature, gravity, radiation, a.o.) and which investigates the effects of branches of the sciences and technology with the aim to determine the changes which technical progress encompasses and how these changes have a feedback on man.*

Here we deal with two kinds of beings of different levels, the Bio Area and the Human Area. How can they be "put together" on this level of the Bio Area 4?
It may be recognized that none of the 4 first areas excludes the human intervention. In fact the sequence of object areas are displayed here in Fig.2 in a rectangular form for the convenience of the onlooker. It ought to be arranged in a circular and thereafter in a spiral form, as of course man cannot be excluded from the very beginning of the subject fields and with each column man's interference intensifies with what precedes his coming into existence in the evolutionary sequence. The first groups contain the more general and theoretical approaches, such as consideration, description and investigation of the existing phenomena, as included in columns 1-6, thereafter only are placed the production and utilization activities as can be seen in columns 7-8 and in the synthesizing column 9 man himself is included as 'object', e.g. under 39 Geography and here under 49 too.

The same holds, however, also for the next level of subdivision, that is, when the systematifier is applied in each subject group; no matter at what point of the system, the same principle is activated and leads from the general and theoretical via the descriptive to the application and synthesis positions of the system which means that already in the first theoretical and general groups we will encounter these concepts on the next deeper level.
Taking this into consideration we may now proceed to apply the facets of the Systematifier to the subject group 49:

49 Ecology and Environment

491 General and theoretical foundations of ecology and environment
492 Ecosystems research, ecobiology
493 Nature and landscape conservation

494 Environmental problems by natural disasters, protection against catastrophes
495 Human ecology (Environmental damage and degradation by technical progress)
496 Environmental protection measures

497 Environmental engineering
498 Environmental economics
499 Education, organization, information in environmental matters

Now, let us go one step further and see what has been proposed to be covered in each case of the nine subdivisions taking into consideration that *only subject fields* (with their definitions) had been collected, defined and allocated in a project of 1978 which led to the construction of the Information Coding Classification as well as the database with the definitions of some 7000 subject fields. It should be kept in mind that within the past 16 years very much progress has been made in this area which would need to be considered in an eventual updating of the following scheme:

49 Ecology and Environment

491 General and theoretical foundations of ecology and environment
491:398 Ecogeography

492 Ecosystems research
4921:238 Ecophysics
49215 Autecology
49216 Synecology, Biocenology
49217 Production biology
4924/5 Hydrobiology, Waterbiology, Ecohydrology
4924 Limno-Biology
4924:428 Limno-microbiology
4924:438 Limno-botany
4924:448 Limno-zoology
4925 Marine biology, sea-water ecology
4925:428 Marine microbiology
4925:438 Marine botany
4925:448 Marine zoology
4926 Terrestrian ecology, ecopedology, soil ecology
4926:438 Plant ecology, phyto ecology, e. of plants
4926:448 Zoo ecology
4927 Settlement biology (Siedlungsbiologie)
4928 Landscape ecology/biology
4928:3922 Tropical ecology
4928:461 Agricultural ecology (Agrarökologie)
4928:471 Forest ecology

493 Nature and landscape conservation
4931 Landscape description/registration
4933 Landscape preservation
4934 Nature (wildlife) and landscape protection
4936 Landscape formation (Landschaftsgestaltung)
4937 Landscape construction (Landschaftsbau)
4938 Garden and parks architecture

494 Environmental problems: Natural desasters, Protection against catastrophes
4942 Storm, wind and hurricane desasters
        and protection against them
4943 Floods and protection against water
4944 Earthquakes and protection
4945 Drought and measures against it
4946 Parasites and insect plagues
        (Schädlingsbekämpfung)
4947 Burning woods and their countermeasures

495 Human ecology (Environmental damage and degradation by technical progress)
4951:238 Noise pollution
4951:284 Radio ecology
4951:268 Environmental chemistry,
        ecological chemistry
4951:348 Air pollution
4951:352 Water pollution
4951:366 Soil pollution
4951:528 Environmental medicine
4951:618 Social ecology (s. a. Urbanism 676)

496 Environmental protection
4961:155 Environmental planning
4961:558 Environmental psychology
4961:628 Environmental policy
4961:623 Environmental legislation
4961:663 Environmental law

497 Environmental engineering
4971:168 Environmental measurement technology
4971:2378 Noise protection measures
4971:2848 Protection from radiation of
        nuclear power plants
4971:5258 Environmental hygiene
4971:758 Construction biology
4971:7934 Treatment of waste
        (Abfall und Beseitigung)

498 Environmental Economics

499 Education, organization, information in environmental matters
        (Environmental education, e. professions,
        e.institutions, e.data collection, e.docu-
        mentation, e.counselling, e.journalism, etc.)

Fig.4: Draft outline of Ecology and Environment (Subject Group ICC 49)

5. Contents and Multidisciplinary Character of the Environmental Sciences

Practically only after WWII the problems of the progressive destruction of nature and the human environment became more obvious than ever before and subsequently scientists, technologists and finally also administrators and legislators started their work of finding out what could be done about them and the media - in need of reporting on such problems - added their share in creating a respective terminology. One problem after the other appeared and needed solutions, subsequently their concepts needed to be delimited and defined. The many combination fields (combinations of notations) shown in the short and rough outline given above demonstrate this drastically.

In a paper on "Domain Interaction..." which included a typology of the different kinds of cross-disciplinarity (11) I showed how the different ontical levels in the ICC can be used to understand what it means that some subject groups possess a *transdisciplinary* property with respect to many of the groups following them. The methods used in the subject groups of the first row are applied in those of the second row and of the first and the second row in those of the third row and so on. Indeed, in the forth row of the Bio Area we find the application of mathematics and statistics, of systemology, organization, measurement, standardization, and control just as the subfields of physics and chemistry, of energy etc. and especially the subject groups of area 3, air, water, land, with all of their subdivisions thematized also as concepts utilized in the environmental sciences.

The advantage of being able to use the concepts already established in their own systematic environment consists in the fact that they carry with them - so-to-speak - all their pertinent other concepts possibly needed. These would then not have to be established separately. This holds, e.g. for 491:393 Ecogeography, where all geographic concepts are included in this notation and need only be activated for a given case just as this would work with all the other combination concepts contained in the ICC schedule shown above.

On the other hand, as can be seen under 493 and 494 there are socalled direct subdivisions: under 493 (the activity facet) kinds of activities were listed in a helpful sequence and under 494 kinds of desasters and their possible protection measures in the sequence of air, water, land, life.

The facet **491** is still rather empty. It should once include all those general concepts which must be available for combinations elsewhere in the sequence of the facets and list also the theoretical concepts having been developed in the meantime for the environmental sciences.

Under facet **492** we placed the *biologically related ecology field* followed later on (beginning at 495) by the *human-oriented environmental sciences*: One might wonder about this combination of apparently two separate things. But nature is indeed our 'natural' basis and if we stop considering its impact on us and vice versa we indeed destroy the basis from which we live materially. Therefore this facet was considered the main object of concern for this subject group 49, but it should include also in some way a subdivision on man's role and interaction within the ecosystem as we not only need to know about the ecosystems but also to learn from them and to live with them and by all means take care that they will be protected from negative influences as this will have an impact on man's well-being.

The facet **493** concerning "*landscapes*" has formerly been treated under Geography (Landscape-geography). However, man has started to include its observation and protection, even its shaping as an assignment of taking care of his environment. Therefore - although the concept of 'landscape' in itself can be referred to as an object it is placed here with regard to all the activities for its proper functioning as the activity facet of the entire subject group.

At facet **494** we are having *Desasters and Catastrophes* - genuinely the specificity facet and with their contents also the reasons per excellence for the problem of Environmental protection.

The specific problems created by nature are somewhat continued by the problems created by man under **495** *Human Ecology*, a very obvious facet for external combinations with concepts from earlier but also from subject groups occurring 'later' in the scheme.

Under **496** follow the actions of society regarding *Environmental protection* by relevant combination concepts with concepts from fields of the human and socio area 5 and 6 where, e.g. *Environmental Legislation* could probably be combined with many of the preceding concepts and concept combinations considering the mass of legislative actions having been taken in the meantime.

Under **497** follow the *technical activities* to be undertaken, again an obvious case for combinations with concepts from previously elaborated subject groups.

498 *Environmental economics* will have its own subdivisions for the genuine economic problems but will also become a field of combinations regarding the economic branches related to the environment, as e.g. the chemical industry, nuclear power plants, fishing, agriculture, horticulture, forestry, etc.

The last facet **499** contains those concepts which are dealing with the *distribution of the knowledge of the environmental science* through education as well as all the kinds of information activities. It contains also the professional aspects, its organizational and actualizational aspects.
Fig.5 shows the relationships between the Subject Group of Ecology and Environment and the other areas of knowledge.

6. Advantages of Concept Systematization

Introducing this paper we summarized some points why we need to care about order and we found:

> Order will bear fruits
> Creation of order makes happy
> Order is needed for understanding
> Order is a necessity for any retrieval purpose
> Order in our lives and environment - a reason for joy.

The order of our concepts has not only been neglected by many colleagues in our times, it has even been denied as an impossible undertaking, probably mainly in fear of the magnitude of the task involved.

And it can be expected with some certainty that the proposed method of applying the Systematifier in the form of a stereotype sequence of facets and the construction and utilization of an overall scheme comprising all knowledge fields will be judged as unrealistic and unscientific, perhaps even unreasonable. However, whoever will make such a judgement will probably not be the practitioner with his daily load of retrieval problems. He must be one of those 'neigbours' who laughed at the poor farmer who sorted the stones according to their size.

The practicality of the Systematifier has been approved in constructing the ICC and in elaborating already some of its Subject groups down to the narrowest concepts. And it has been applied recently also in structuring the field of Italian literature (12, 13).

Some of the advantages of using this device can be related also to the principles which Fugmann (2) established with respect to an adequate indexing practice. These principles

0 GENERAL FORM CONCEPTS	01 THEORIES, PRINCIPLES	02 OBJECT, COMPONENT	03 ACTIVITY, PROCESS	04 PROPERTY ATTRIBUTE	05 PERSONS OR CONT'D	06 INSTITUTION OR CONT'D	07 TECHNOLOGY & PRODUCTION	08 APPLICATION & DETERMINAT.	09 DISTRIBUTION & SYNTHESIS
1 FORM & STRUCTURE AREA	11 Logic	12 Mathematics	13 Statistics	14 Systemology	15 Organization Science	16 Metrology	17 Cybernetics, Control & Automation	18 Standardization	19 Testing and Monitoring
2 ENERGY & MATTER AREA	21 Mechanics	22 Physics of Matter	23 General and Technical Physics	24 Electronics	25 Physical Chemistry	26 Pure Chemistry	27 Chemical Technology & Engineering	28 Energy Science and Technology	29 Electrical Engineering
3 COSMO & GEO-AREA	31 Astronomy & Astrophysics	32 Astronautics & Space Research	33 Basic Geosciences	34 Atmospheric Sciences & Technology	35 Hydrospheric & Oceanol. Sci & Technology	36 Geological Sciences	37 Mining	38 Materials Science & Technology	39 Geography
4 BIO-AREA	41 Basic biological Sciences	42 Microbiology and Cultivation	43 Plant Biology and Cultivation	44 Animal Biology and Breeding	45 Veterinary Sciences	46 Agriculture & Horticulture	47 Forestry & Wood Sci. & Technology	48 Food Science & Technology	49 Ecology and Environment
5 HUMAN AREA	51 Human Biology	52 Health and Theoretical Medicine	53 Pathology and Practical Medicine	54 Clinical Medicine & Cure	55 Psychology	56 Education	57 Profession Sci., Labor, Leisure	58 Sport Science and Sports	59 Household and Home Life
6 SOCIO AREA	61 Sociology	62 State and Politics	63 Public Administration	64 Money and Finances	65 Social Aid, Social Politics	66 Law	67 Area Planning, Urbanism	68 Military Science and Technology	69 History Science and History
7 ECONOMICS & TECHNOLOGY AREA	71 General and National Economics	72 Business Economics	73 Technology in general	74 Mechanical & Precision Engineering	75 Building	76 Commodity Science & Technology	77 Vehicle Science and Technology	78 Transportation Technology & Services	79 Utilities and Service Economics
8 SCIENCE & INFORMATION AREA	81 Science of Science	82 Information Science	83 Informatics, computer science	84 Information in general	85 Communicat. Science	86 Mass Communication	87 Printing and Publishing	88 Communication Engineering	89 Semiotics
9 CULTURE AREA	91 Language and Linguistics	92 Literature and Philology	93 Music and Musicology	94 Fine Arts	95 Performing Arts	96 Culture Sciences, narrower sense	97 Philosophy	98 Religion and Secret Teachings	99 Christian Religion

Fig.5: Information Coding Classification. (c) 1982, rev.1992. I.Dahlberg
Subject Group 49: Ecology and Environment
and its relationship to other knowledge fields.

are also valid as postulates for the systematization of knowledge fields by a Systematifier, given here in a somewhat different sequence:

1. *Order:* The arrangement of facets by a Systematifier is an order creating activity.

2. *Defineability:* The order of concepts according to facets is based on their defineability with respect to their categories: entities/objects, properties, activities and their derived concepts.

3. *Degrees of order:* The systematifier can be used on any level of subdivision. It implicitly contains an ordering scheme which raises the degree of order of any collection of concepts arranged subsequently.

4. *Predictability.* The single facets of the Systematifier are arranged in such a way that a user - after a short introduction - will recognize immediately at which element position in the system which kinds of concepts and subjects may be located.

5. *Precision:* The arrangement of the facets is done in such a way that the concepts can be combined with each other internally within one subject group and externally between different subject groups. Rules can be elaborated which unambiguously facilitate the combination of concepts in all necessary *precision.*

We would like to add two further advantages: The Systematifier acts like a "scaffold" (9) in structuring a knowledge field and therefore it can be regarded by its

6. *Constructor friendliness:* The given outline facilitates the construction of any concept system in a subject field, and

7. *User friendliness:* Because of its predefined system positions a user should find easily the position of the concepts he is looking for (see also the postulate of predictability above).

7. Future Prospects

The more catastrophes are occurring on this our globe at present the more people become conscious of the necessity to do something to save their own direct and larger environment. Therefore the environmental sciences will probably soon grow at a very fast pace in two applications: Data documentation and literature documentation. For both cases we need to have prepared at least three items:

1) a clarified terminology with their definitions for the concepts in question

2) a most suitable concept system offering the system positions necessary to capture all the relevant concepts

3) a list of all those problems which might become relevant and which need terms, definitions and a position in a concept system to meet possible future developments.

The recently published "Encyclopedia of World Problems and Human Potential" (14) contains also a long list of problems with respect to the environment. Especially under the concept of "Environmental Hazards" we may find many entries which would relate to the item 3). Hazards need countermeasures fast and it is therefore recommended to join forces in some way and collaborate towards being prepared with the 3 items listed above, which are not meant to be elaborated for their own sake but for application in either computerized information retrieval systems or in new kinds of reference books which, eventually, may also be computerized in order that fast solutions will be available for whatever need may come up suddenly and inexpectedly.

Note:
1 This story is not contained in the New Testament but in Vol.1, Chapter 92, of the Great Gospel of John, revealed through the Grace of the Lord to Jacob Lorber (1800-1864). The 11 volumes of this Great Gospel (7th ed.) are available from the Lorber Verlag, Hindenburgstr.3, D-74321 Bietigheim)

References:
(1)   Fugmann, R.: Ordnung - oberstes Gebot in der Dokumentation. Nachr.Dok. 13(1962)No.3, p.120-132

(2)   Fugmann, R.: Theoretische Grundlagen der Indexierungspraxis. Frankfurt: INDEKS Verlag 1992. XVI,325p. = Fortschritte in der Wissensorganisation, Bd.1

(3)   Fugmann, R.: Subject analysis and indexing. Theoretical foundations and practical advice. Frankfurt: INDEKS Verlag 1993. XVI,256p. = Textbooks for Knowledge Organization, Vol.1

(4)   Dahlberg, I.: A referent-oriented concept theory for INTERCONCEPT. Int.Classif. 8(1981)No.1, p.142-151

(5)   Dahlberg, I.: Concept and definition theory. In: Classification Theory in the Computer Age: Conversations across the disciplines. Proc.from the Conf., Albany, NY, Nov.18-19, 1988. Albany, NY: School of Inform.Sci.& Policy, SUNY 1989. p.12-24

(6)   Normenausschuß Klassifikation (NAK) im DIN Deutsches Institut für Normung e.V.: DIN 32705. Jan. 1987. Klassifikationssysteme. Erstellung und Weiterentwicklung von Klassifikationssystemen. Berlin: Beuth Verlag 1987. 12 p.

(7)   Dahlberg, I.: Grundlagen universaler Wissensordnung. München: Verlag Dokumentation K.G.Saur 1974. XVIII,366p. = DGD Schriftenreihe Bd.3

(8)   Dahlberg, I.: Information Coding Classification - principles, structure and application possibilities. Int.Classif.9(1982)No.2, p.87-93. Also in: Classification Systems and Thesauri, 1950-1982. Frankfurt: INDEKS Verlag 1982. p.107-132 = Int.Classif.& Indexing Bibliography, Vol.1

(9)   Judge, A.J.N.: Visualization: Structural outliners and conceptual scaffolding. In: Encyclopedia of World Problems and Human Potential. 4th ed., Vol.2. München: jK.G.Saur Verlag 1994. p.535-536

(10) Dahlberg, I.: Ontical structures and universal classification. Bangalore: Sarada Ranganathan Endowment for Library Science 1978. 76p.

(11) Dahlberg, I.: Domain interaction: Theory and practice. In: Albrechtsen, H. et al: Knowledge Organization and Quality Management. Proc.3rd Int. ISKO Conf., 20-24 June 1994, Copenhagen, Denmark. Frankfurt: INDEKS Verlag 1994. p.60-71

(12) Aschero, B., Negrini, G. et al: Systematifier: a guide for the systematization of Italian literature. In: Meder, N. et al: Konstruktion und Retrieval von Wissen. 3.Tagg.d.Dt.ISKO Sektion, Weilburg, 27-29 Oct.1993. Frankfurt: INDEKS Verlag 1994. (In preparation)

(13) Massimiliano, G., Negrini, G.: A tool to guide the logical process of conceptual structuring. In: Albrechtsen, H. et al: Knowledge Organization and Quality Management, Proc.3rd Int.ISKO Conf., 20-24 June 1994, Copenhagen, Denmark. Frankfurt: INDEKS Verlag 1994. p.342-349

(14) Encyclopedia of World Problems and Human Potential. 4th ed., Vols.1 and 2. Edited by Union of International Associations, Brussels. München etc.: K.G.Saur Verlag 1994. 1258+929p. (here Vol.2, p.1023-1024).

Otto Nacke, Tobias Krull
Institut für Veritologie, Bad Salzuflen, Germany

Structures of truth

When we proposed this paper, we were expecting the domain 'truth' to be as much represented in environmental research literature as it is in medicine, psychology and chemistry. We were wrong, among the 25 388 publications scientometrically analyzed by Krull we could only find 17 dealing with truth problems. This was not enough for a serious scientometrical study of this subject.

1. Properties of the concept 'truth'
1.1 Definitions and Analysis
What is 'truth'? Here are 14 definitions.

1   Truth or untruth are solely in a proposition. A proposition is true, if it is in correspondence with its object.
2   Truth is the correspondence of cognition and an object.
3   A statement is truely accurate if and when the facts in reality act, as claimed in the statement.
4   The truth of a statement can be defined as its correspondence with objective reality.
5   A statement is true, if it causes expectations which can be confirmed by every normal fellow men.
6   Truth is a feature of statements and thoughts.
7   Truth is the correspondence of thinking with its item.
8   True is, what useful is.
9   Truth is a philosophic category which reflects the adequacy of the perception with its real object.
10 Practice is the criterion of truth.
11 A statement, an assertion or a report are true if they correspond to the facts in reality.
12 Truth is the correspondence, the appropriateness, of the proposition and its object.
13 Truth is a relation between an idea and its object.
14 Truth is the logical consistency of a proposition within the relevant system of propositions.

In summing up: in dealing with 'truth' we are dealing with a two-sided relation. Anything could be in a 'truth'-relation with any other thing.
An analysis of the 14 definitions is shown in the following table.

	Front link	Back link	Relator
1	proposition	object	congruence
3	statement	the facts in reality	facts behave as assertions
4	statement	objective reality	congruence
5	statement	expectations	expectations which can be confirmed by every normal fellow men .
14	statement	relevant system of statements	consistency
11	statement, report, assertion	facts of reality	corresponding
6	statement and thought		
7	thinking	object (of thinking)	congruence
2	recognition	object of cognisance	congruence
9	recognition	objective real object of cognisance	reflection
13	idea	object	relation
10			Practice is the criterion of truth
8			what useful is
12	proposition	the object	validity, adequacy

What this is, what is said to be true (this means, the front link of the truth relation), originates from three areas : language, thinking and existence. Therefore it is linguistically, psychologically or ontologically defined. In addition to this simple assignation there are further hybrid forms.

1.2 Dimensions of the front link

Language	thinking	existence	hybrid forms
linguistic	psychological	ontological	
proposition statement	thinking	idea	propositions and
report assertion	cognition		thoughts

The back link is defined in two forms, depending on the world view the definer assumes: objectivistic or subjectivistic.
If the definer believes that what we "perceive" is a copy of reality, he defines in an objectivistic way. If he does not believe this, he has the back link to be defined as 'object of cognition' or 'object of thinking'.

1.3 Dimensions of the back link

objectivistic	subjectivistic
objective reality	object of cognition
facts the reality	object (of thinking)
objective real object of cognisance	object of cognisance
	object

The relator could work either formally or operationally. In the operational case, an operation, a procedure (something that is executable), is named. This in order to determine, if front link and back link are linked by the relation 'true'.
For example referring to the definition:

"A statement is true, if it causes expectations which can be confirmed by every normal fellow men ."

the procedure is: asking "normal fellows" for their expectations.

1.4 Dimensions of the relator

formal - rational	operationally - empirical
reflection	consulting normal fellow men
congruence	facts behave as assertions
consistency	practice is the criterion of truth
adequacy	what useful is
trueness	

In all the above definitions the criterion 'true' is treated as a dichotomy. That means that solely the 'true' and 'not true' criterion of the classical logic are allowed. This dichotomy is certainly feasible without contradiction, but only very roughly.
For example: a boy of 11 years, whose marks at school are "very good" in some fields and bad in others.
If he is pretended to be a girl, then this is surely unequivocally wrong, here there is no "third".
If someone affirms he is 12 and another tells you he is 61, it is obvious that between these two statements there is a difference in the "range of truth".
if his parents tell, he is said to be "very good" at school, then this is true on the one hand and false on the other.
This shows that truth is scaleable.

The following chart shows an example.

1.5 Gradual steps of truth

attribute	linguistic formulation	example
alternative	either --- or --- right	the gross national product raised in Spain
constantly	more or less right	the gross national product raised in Spain by 3%
particularly	part --- part --- right	the gross national product raised in the countries of the EC

2. Disciplinary Aspects of Truth

The problem of truth is a multidisciplinary problem. The six disciplines most involved here are: philosophy, logic, psychology, linguistic, statistics and jurisprudence.

2.1 Philosophy
Philosophy, the main goal of which is to recognize the real principles of things, is intensively dealing with truth problems. Four main domains are interested in these crucial questions : ontology, ethics, epistemology and the theory of science.

2.1.1 Ontology
Ontology, which investigates the nature of existence, is asking for the true being. According to Platonic and scholastic understanding, truth is a property that is attached to the things as such. 'Truth' stands for 'reality', both concepts were considered equal.

2.1.2 Ethics
The ethical aspects of truth belong to everyday life problems.
But, do we have to tell the truth to:
> Deadly sick persons?
> Contented fools?
> Dangerous fanatics?

The ethical investigation of such problems has to consider the relativity of judgements, different customs could lead to different judgements. How to solve these problems is relative to the principles you use as your ethical reference.

2.1.3 Epistemology
If truth is defined as a relation between reality and its reflection, then the main truth-question of epistemology is: "how far can we be sure to recognize reality?".
When Kant says: "we recognise in things only what we have put into them ", he speaks in favor of the *rationalism* of epistemology. This is in opposition to *empiricism*, meaning that only experience leads to knowledge. "There is no innate knowledge".

2.1.4 Theory of Science
Theory of science is mainly natural science theory. Throug the fact that the main goal of natural science is to find nature's laws - and these laws are universal laws - the main problem of the theory of science is the verifiability of the truth of empirical statements. According to Popper these are not verifiable, only falsifiable. Therefore only falsifiable statements are allowed in science.

2.2 Logic

Logic investigates the formal conditions of truth. The principles of this discipline had been developed by Aristotle in the 'classical logic' completed in the last century by 'symbolic or mathematical logic'.
For the treatment of truth problems, the experience shows that classical logic gives better results than formal logic. The four main domains of classical logic are dealing with important truth problems: logic of concepts, logic of relations, logic of propositions and logic of deduction.

2.2.1 Logic of concepts
The problem we have to face here is the fuzziness of concepts occurring quite often. We have to distinguish three types.

A) Unsharp concepts

The borders of concepts are not always sharp:
Where does orange stop and red begin? When is a car "used", after 10, 100 or 1000 miles? When can you say that an injury is light? Nobody can say where the mountain stops and the valley begins. When you reach those limits a "truth-space" appears, where you have the choice of using a word or its "border-word". For example, red, or orange, or not-red, or not-orange could be used with the same right to truth in the "truth-space". This opposes the logical axiom of contradiction that something cannot "have a property" and "not have a property" simultaneously.

B) Homonyms

Homonyms have different meanings but the same form, for example the word tank stays for a gasoline reserve but also for a weapon.

C) Indeterminate concepts

In opposition to the previous case these concepts have in fact only one main meaning, but this meaning allows very different interpretations. What does it mean: "this is healthy"?
-    Does it reduce the number of sick people? If so the economical crisis is healthy.
     The company doctor knows: is the company sick then the worker gets healthy.
-    Does it increase physical efficiency? If so, doping should be a civil duty.
-    Does it increase the live expectancy? If so, environment damages are healthy. Because in the last 50 years, both life expectation and environment damages increased impressively .

Who can be considered a Christian ?
     The one who is baptized?
     The one who pays taxes for his church (in Germany for example)?
     The one who believes in what God says?
     Or at last the one who tries to live in accordance with God's laws ?
One concept and quite different interpretations. This becomes especially clear, if the different interpretations are taken as definitions for statistical counting.

2.2.2 Logic of relations

Truth problems can also occur with an absolute use of relative statements, and by a logically false use of relations.

A) Absolute use of relative statements

In the eyes of some fundamentalist, killing unfaithful people is a good act. For people not being fundamentalists, this is a crime.
Some insecticides are bad for the environment, but good for the economy.
If a professional pianist looses his little finger it is a 100 percent injury, by contrast it is a 0 percent injury for a consultant.

B) False use of relations

We have a good example, in the relation "same". It is based on the acceptance that things that have the same behaviour in some cases, have another behaviour in other cases.
All farmers know, that when you introduce a foreign pig in a group it will be killed over night by the others. This foreign hate could be considered "natural" and this is real in this case. But we are not pigs, we are humans and we are able to overcome our instincts. Racism is not to be considered as natural for humans.

2.2.3 Logic of propositions

Propositions are statements which can be true or false. Most of the time they have the grammatical form of a sentence, their designatum is the fact. Near to it are two other sentence forms:
     Commands, which are useful or not useful.
     Fictions, that are possible or not possible.

The truth problem of statements can be sorted into three groups. Problems occurring with:
     The designatum, the object of the statement.

The method of verification.
The statement itself.

A) Designatum
The problem is here to evaluate at which degree the truth of the statement is determinable. For example:
Is Bob a boy?
Has Bob ever been to Berlin?
Has Bob tried to cheat when playing Monopoly or was he just inattentive?

In the first case you can verify the proposition by looking into his trousers,
in the second by looking into his diary (if he has one), and
in the third case by looking into his head.
As you see the truth determinability of these sentences are principally different.

B) Method
For this point the main problem is what is admitted pragmatically as true:
something that was admitted to be surely real or,
something that was not considered as false or,
something that was said in the Bible, or
something that could be verified empirically, or, or, or...?

C) The proposition itself
In this case, the number of truth problems is so great, that only one should be given as an example:
The confusion between the different logical forms of grammatical sentences:
A definition can be:
- A proposition (which is right or wrong) when it says how a word **is** used.
- A command (which is useful or not useful) when it says how a word **should be** used.
- A fiction (which is possible or not possible) when it says how a word possibly **could be** used.

2.2.4 Logic of deduction
Deduction is the formal creation of statements based on other statements.
The Logic of deduction examines rules: if you act against them, deduction failures will occur.

Here are some examples:

A) Going against the rules
From "if A then B" you cannot conclude "If B then A".
"When it rains the street is wet" does not lead to:
"the street is wet so it has been raining".

B) Use of false rules
For example: the deduction from naturally to healthy.

C) Wrong order of argumentation
For example: the circular reasoning, which proves the truth of a proposition by the contents of the proposition itself.
For example: deducing that a special law is right because this law is in a code; and - all laws that are in the code are right.

2.3 Psychology.

We are dealing here with four main points: sense mistakes, suggestion, preconception, thinking mistakes.

A) Sense mistakes
They are due to bad translation of the messages your senses send to you. You are not able to recognise the real situation. As an example think about optical illusions.

B) Suggestion

We must make a difference between self-suggestion and external suggestion. Both are ruled by the object and the personality. The more affective the person the more suggestive the effect. Words like "a non-genuine refugee" are very suggestive.

Here we can also speak about the placebo effect, which shows that some people (so-called 'placebo reactors' by psychiatrists) get healed just because they knew they where given the right medicine. These effects are related to the suggestibility of patients.

C) Preconception

Decision backtracking is very current in our behaviour, it is not useful to redo the same reflection every time you face a known situation. We are all using our preconceptions, we call them "our experience". Only false preconceptions are a source of failure.

We have individual and collective preconceptions, the second ones allow us to create social groups leading to human closeness in populations which have the same collective preconceptions.

D) Thinking mistakes

We can identify two separate groups, formal and informal mistakes.

The first group contains mostly logical failures in deductions. They occur when one breaks the rules of thinking.

Informal thinking mistakes are much more varying. As an example we have the well studied argumentation failures:

Argumentum ad baculum: appeal to the "stick"
Argumentum ad populum: appeal to the people
Argumentum ad misericordium: appeal to pity
Argumentum ad hominem: argument against the person
Argumentum ad vericundiam: appeal to authority
Argumentum ad ignorantiam: appeal to ignorance

2.4 Jurisprudence

We can find two major relations between law and truth.

On the one hand law is based on truth. It determines for itself what must be considered as true.

On the other hand law tells you where and when you have the right to hide or tell the truth.

We can show here three relations between truth and law.

A) Professional discretion

It is obvious that doctors, lawyers and clergymen must not reveal private information. This discretion is very strictly protected by the law.

B) The right to remain silent

Labour legislation gives us here plenty of examples. When applying for a job you have the right not to reveal all information you are asked for, you even have the right to lye. A witness has also the right not to come to a trial.

C) Freedom of speech

In all democratic countries the freedom of speech is a constitutional right. In Germany it is the fifth article of the constitution: "Everybody has the right to express and diffuse his opinion with various means, by speech, text or images. The right to get freely in contact with any source of information is also granted. The press and all the other means of expression like radiocomunication and film are also free. There will be no censure of any kind."

2.5 Mathematics (Statistics)

Statistics is the science of investigating statistical facts. These facts are a significant part of our life, but most of the time we do not notice them.

In this supermarket things are cheaper.	Not just x percent of the products.
If it rains now, all my corn will be damaged.	Not just x percent of the whole corn.
Letters arrive within one day at their destination.	Not just x percent of all letters.

Statistical statements are treated in three steps:
Gathering of data, mathematical treatment, interpretation and representation of the results.

A) Gathering of data
That the quality of a statistic study depends on the quality of the data is an obvious thing. Here we have a source of many mistakes.

While investigating the effects of an analgesic:
how can you distinguish low pain from middle and strong pain?
In criminalistics, how high are the hidden figures of abused children?

B) Mathematical treatment
The following mistakes could occur:
Use made of the Chiquadrat test on data that are not normally grouped.
The way of grouping data in classes biasing the original material.
The use of statistical methods on data the quality or quantity
of which are not adapted to these methods.

C) Interpreting and representing data
The mistakes mostly encountered are:

C1) Bad interpretation of correlation:
The fact that a positive correlation exists is a "needed condition" but not a "sufficient condition" for a causal relationship.
The length of your left arm is strongly correlated to the one of your right arm, but here there is no causal relationship.

C2) False precision
7 on 17 can be represented by 24 percent, it would be senseless representing this with 24, 285 percent.

C3) Non-allowed comparison between measured values
For example: radioratings are rather a negative than a positive indicator for the "cultural content" of a TV program.

2.6 Linguistics

Three domains where linguistics and veritology overlap should be named here: terminology research, dialectics, text analysis.

2.6.1 Terminology research
One main problem here is the professional vocabulary (medicine, technology etc.). In the moment it is popularized it looses its real meaning. These problems are often encountered in the domain of lexicology that is aimed at setting up dictionaries. We face here a national and international problem.

2.6.2 Dialectics
What is investigated here is the art of intellectual dispute, among which one may find psychological and pure tactical sides of persuasion. It deals with permitted and non-permitted techniques of argumentation.

2.6.3 Text analysis
With this method one tries to detect "behind" the written text what the author really aimed at.

3. Conclusion

At the end my of my paper I want to come back to the astonishing beginning: only 17 of the 25,388 analyzed publications treat the problems of truth and error.
How to explain this probably unique scientific phenomenon?

Does ecology not pose veritological problems?
The opposite is right. Go simply across the disciplines listed above and you will easily be able to construct examples for veritological problems of ecology. This begins already with terminology. Depending on the definition of "toxic" you can assume, bread is poison and prussic acid is no poison.
Ecological problems are to a great part statistical problems. If these problems combine with the difficulties of causal analysis, then the difficulty of finding the true cause of a damage increases exponentially.
The extremely difficult problems of ecological carcinogenesis' statistics results in a great number of articles in medical periodicals. A few years ago a important medical congress had dealt with these questions as its only theme. Within 74 ecological periodicals studied by Krull, not a single one treated this complex question of epidemiological investigation of ecological carcinogenesis!

Hence it is not the absence of ecological problems that explains the absence of veritological research
What is it than?

Three facts seem to explain this phenomenon:

1. Ecology, compared with most of the other established sciences, such as medicine, astronomy, mathematics, philosophy, is still a young science. It is still in the conquest phase of its terrain. The attitude of self-criticism, of doubting and scepticism is, however, something more linked to the wisdom of the aged.
Hence we can still raise hope for a veritology of ecology in the future . We will let it have some time to become a bit older. Perhaps this paper gives a little kick in this direction?

2. Ecology is not only a science, but often also a confession. Only the true belief may be preached! Doubters, criticizers and sceptics are to be delivered to the stake. And who wants this?

3. Ecology and politics are closely networked with each other. This is both positive and negative for ecology. Politicians want to be made the champions of ecological research promotion. But the politician that declares the aim of his ecological politics is the promotion of critical research in this field has still to be born. Therefore, without politics no promotion, without promotion no money and without money no research.

Note

1 The 14 definitions (defined by the authors) have mostly been taken from philosophical dictionaries, lexicons and textbooks. It does not seem to matter to list the references, however, this could be done upon request. The author is ready to help in such cases.

Maria Domokos, Endre Dudich
Budapest, Hungary

On the Possible Use of Organization Levels and Time/Space Parameters in the Classification of Natural and Man-Made Objects

Abstract

1. A prerequisite of any classification of beings constituting the (human) environment is to define what "environment" means (or should mean). This can be done on subsequent levels of abstraction: natural science - mathematics - ontology (metaphysics) - natural theology.

When we intend to classify "natural and man-made objects" we are obviously on level (1), with a possible outlook for help on level (2).
At any rate "environment" is a relational concept, and we have to deal with a dynamic equilibrium of (complex) systems.

2. These (systems) can reasonably be ordered according to organization levels, on the basis of their structural and functional complexity. The organization levels can conveniently be subdivided into two series, corresponding to the "non-living" and "living" world. The first (anorganic) one goes from the physical camp to the metagalaxy, while the second (organic) one from the organic macromolecules to the biosphere. In this scheme, "man" is a living organism, a "human community" is a population, and "mankind" as such is a species, one of those that constitute the biosphere. Accordingly, it is important to specify each time whether we are speaking about the environment of an individual, a particular group, or the totality of human beings.

3. Another possible approach is to order beings according to their time and space parameters (roughly speaking, size and duration, or "velocity" of motion and existence). This has been done by Prof. Szadeczky-Kardoss, with the surprising result that in the double logarithmic time/space (sec/cm) diagram the "stable" objects and phenomena of our Universe appear arranged in four bands: nuclear, electromagnetic, mechanical and chemical (inclucing geological and biological). This succession of enumeration is also that of increasing complexity.
For the simpler systems a classification based only on the t/s parameters might be fairly appropriate, and it would be extremely easy to handle it mathematically.
For the complex (geological, chemical, biological) systems a hierarchical system of organization levels has to be superimposed on the previous one.

4. The resulting technical problem is to work out an adequate coding system for the two interactive series of organization levels. This has to be flexible enough to reflect also their constantly changing dynamic equilibrium. (No static classification would be acceptable.)

We have been strongly tempted to undertake the elaboration of an apparently very simple binary code system, expressing the (eventually multiple) superposition of sets elements and subsystems. The first trial dates back to 1974. The present lecture produces a more developed version, as a basis of discussion and further elaboration.

Wim W. de MES
Rapporteur Unesco/IHP, The Netherlands

Access to Online Databases dealing with Water and the Environment

Abstract: Online databases in the Water and Environment sector form part of the field of attention of the Working Group on Information Management working under the aegis of the Unesco, Division of Water Sciences. One of the problems in the use of these databases, is the lack of consistent and unified indexing: retrieval strongly depends on words from the titles (spelling and language problems), source information, and abstracts. This problem origins in part from the fact that most public online databases in this sector have a long history as either card catalogues or abstract journals from specific institutions, for a limited user group, or for a certain market sector, so the indexing methods differ widely. Some databases have a well developed thesaurus, some use classifications or subject headings, others use list of controlled terms or keywords. Some small comparative studies have been done to quantify this problem, some of the results will be presented, but a solution is difficult. Many hosts have developed "user-friendly" menu-driven searching, also CD-ROM's have menu-driven searching with extensive direct help, but still the use of the right search terms remains a problem. Meanwhile full-text systems are arriving, pronouncing the use of "natural language" terms. Expert systems might bring a solution, but only in small fields of interest.
The Unesco Working Group is now involved in a feasibility study on the development of an electronic multilingual glossary on Water Resources, which should be developed into a multilingual thesaurus. The idea is to use existing sources as much as possible, so at the start an inventory will be made. Eventual progress on this project will be reported.

0. Introduction

Online databases in the Water and Environment sector form part of the field of attention of the Working Group on Information Management working under the aegis of the Unesco, Division of Water Sciences, within the International Hydrological Programme (IHP-IV 1990-1995). The use of this databases, however, is obstructed by a number of facts:

a - poor telecommunication services in parts of the world mostly in need of information

b - different search languages for each host

c - the lack of consistent and unified indexing: retrieval strongly depends on words from the titles (spelling and language problems), source information, and abstracts

d - coverage of the databases, white spots and overlap, which database is where?

Problem (a) is gradually solved by the growing use and possibilities of Internet, meant primarily for E-mail, but most hosts are connected also, and the growth of the number of CD-ROM products. Problem (b) is solved by menu-driven at the price of slower response, less accurate results and higher online costs. Problem (c) origins from the fact that most public online databases in the Water and Environment sector have a long history as either card catalogues or abstract journals from specific institutions, for a limited user group, or for a certain market sector, so the indexing methods differ widely. Some databases have a well developed thesaurus, some use classifications or subject headings, others use list of controlled terms or keywords (and call these lists their "thesaurus").

Knowledge Organization in Subject Areas, INDEKS Verlag, Vol.1(1994)p.104-111

Some small comparative studies have been done to quantify the last problem (d), some of the results have been presented earlier [ref 2], but a solution is difficult.

Also many hosts have developed "user-friendly" menu-driven searching, but still the use of the right search terms remains a problem. Meanwhile full-text systems are arriving, pronouncing the use of "natural language" terms. Expert systems might bring a solution, but only in small fields of interest.

The Unesco Working Group is now involved in a feasibility study, initiated by the Subcommittee for Water Resources of the UN/ACC (Administrative Coordinating Committee), on the development of an electronic multilingual glossary on Water Resources, which should be developed into a multilingual thesaurus. Such a glossary could be useful to standardize terms and to overcome language problems or finally develop into an interface or help-database on CD-ROM, using CDS/Isis and the Heurisko interface. The idea is to use existing sources as much as possible, so at the start an inventory will be made.

1. The UNESCO/IHP Project IV-M-2-1 and M-2-2 (formerly III.17.1)

The project IV-M-2 is entitled *Scientific and technical water related information and documentation systems*. The general objectives are to promote, provide methodologies and to assist in the development and establishment of such systems. The main goals of project IV-M-2-2 were: to inform people on the existence and possibilities of the online databases, provide guidance to end-users and promote the use, for instance to incorporate it in post-graduate courses. The Working Group started in the early eighties.

A report has been prepared containing an overview and details of existing databases in this field also including CD-ROM. General information sources on "databases" only give large quantities of general information. The Unesco Working Group felt that there was a strong need for a specialised guide in the water field. The feeling was also expressed by many people working in the information management field. So it was decided to produce such a guide.[ref 6]

2 "Online" databases.

2.1 Short history

Almost all bibliographic databases started as abstract journals or even simple card files. Many abstract journals were at first meant for a relatively limited user group, especially in the water field. This meant that the scope of this journal was strongly determined by the specific needs of that specialised group. And as there is in general a strong overlap between the scope of, for instance, research institutes, or science fields (like hydrology, water supply and sanitation, hydraulic engineering etc.), there was also a strong overlap in the abstract journals.

The production of abstract journals has been strongly influenced by the progress in printing techniques. When computer typesetting came up, the contents of the abstract journal was available in "digital" form, which made the production easier and contributed to the more efficient production, for instance, it became much easier to produce monthly and annual indexes.

Especially in the fields where money, at that time, seemed to be no problem, like the chemical industry and aerospace, the digital form developed soon into databases. Also in this field there was a strong need to keep up to date, especially after the launch of the first Russian satellite. So the NASA/DOE Recon system was one of the first text retrieval software packages to be used for online "public" access, and it is still the ancestor of two important hosts in the technical field, Dialog and ESA/IRS.

When the prices of computer facilities went down and the possibilities of storage and processing grew strongly, more and more abstract journals became also available on line. For instance Aqualine came online in 1974 and Water Resources Abstracts a little earlier, in 1972 and fully available also in 1974.

As water is important in many disciplines, water information can be found in databases covering many different subjects. Hydrologists are interested in rainfall-runoff relations, geologists like to know about the groundwater, biologists are interested in everything living in, on, around surface waters, and so on and so on. The result is that bibliographical information on "water" can be found in a large number of databases.

The technological development brought again new media, which tend to have again more possiblities for digital storage at an even lower price. So now many databases are becoming available in the form of CD-ROM (Compact Disc Read Only Memory). The guide mentioned [ref 6], also includes these databases. CD-ROM's have a large storage capacity (600 Mb), use standardized technology, and they are relatively insensitive to environmental conditions. The costs to produce a CD-ROM are rapidly declining. In the last few years prices of CD-ROM drives have been dramatically reduced.

In the "water" field a limited number of databases is available, but much information can also be found in water-related or "mega-databases". On the other hand a number of database producers is considering to publish their databases on CD-ROM, like International Civil Engineering Abstracts, which has never been available online.

2.2 Hosts

To make the picture of online searching a little more clear, a few words on hosts. Hosts, earlier often called "spinners", are exploiting computer-systems, which provide access to databases and offer additional services on a COMMERCIAL basis.

Each host has its own SEARCH language, connected with the software used. So in each host, COMMANDS and SYNTAXES are different.

Some host-services have taken special measures to facilitate access and use of information in the databases for end-users. (non-skilled searchers). Dialog (located in the USA) introduced the Knowledge Index, using a limited number of databases and a simplified search language. Dialog also started with the Business- and Medical Connection, and BRS (also in the USA) has its AfterDark and BRKTHRU (USA only) facilities which all operate with a menu-driven system. Also ESA/IRS (in Italy) can be searched using menu's. ESA/IRS is still experimenting on searching using natural language, as well as investigating the possibilities of creating an "intelligent interface" .

The use of multiple databases with one and the same search-strategy, is facilitated.

Onesearch at Dialog searches the selected databases one after another. Cluster searching at ESA/IRS searches the selected files simultaneously and should give a stronger reduction in search/connect time. Dialog was the first to add the possibility to remove duplicates from the search results, after executing a "Onesearch" on multiple databases. This "deduplication" is now also available on many other hosts.

Charges are generally based on connect time and number of abstracts printed.
ESA/IRS made the step to change the charge-policy completely; the costs for connect time are very low, a fixed price for each database entered is charged (between $6 to $12) and only if you really take abstracts or parts of it from a database you have to pay. It is stated that in this case the "taxi-meter" effect is absent, which will result in better searches, better results, and better use of all the possibilities of the search facilities. This example has only been followed by Australis, an Australian host.

There is a tendency of concentration, some hosts are merging (Dialog took over DataStar (Switzerland) and Questel (French Telecom) took over Orbit Infoline.

2.3 Databases in detail.

For the purpose of this paper a very general division of databases is made, i.e.
- "Water" databases
- "Water-related" databases
- "Mega-databases"

2.3.1 "Water"-databases

The databases closely related to water are spread over a number of hosts.

On Dialog we find the databases Aquaculture (closed file since 1984), Aquatic Sciences and Fisheries Abstracts (ASFA, produced with FAO), Water Resources Abstracts, Waternet (produced by the American Water Works Association(AWWA)), and FLUIDEX (produced by BHRA, but now by Elsevier). ESA/IRS has recently added ASFA next to the databases EAUDOC (AFEE), Delft Hydro (no longer updated, last update 86/12 and recently put offline), and FLUIDEX. Orbit-Infoline had exclusively Aqualine and had Waterlit. Waterlit is now available on CD-ROM only. FIZ-Technik has brought Hydroline publicly on-line as a part of the Geoline database, this database is also on STN. STN recently added Civil Engineering Database, containing the publications of the American Society of Civil Engineers. ESA/IRS loaded in the last few years also Aqualine, and Water Resources Research.

Each database has its pro's and con's. Water Resources Abstracts is very complete in coverage, but is rather slow in updates, Waterlit has no abstracts, EAUDOC has brief abstracts only in the most recent years. Some are indexed manually, some are machine-indexed, or not at all. EAUDOC has French controlled terms, most other databases have English or US-English controlled terms, except Fluidex which has none. In addition, many databases also use some kind of classification. A general problem is the retrievability of geographic areas. Most databases only have the formal terms (from title or abstract) to retrieve the references, or have very general index terms.

An overview of a number of retrievable fields, and their differences, for a number of databases is given.

Database	Indexing	Index-language	geographic indexing	Abstract
EAUDOC (AFEE)	manual	French	CT=(large areas)	French
Aqualine	automatic/manual	English		English
Delft Hydro	manual	English	CT=, CC=	original language English
Compendex	manual	English	UT or CT=(large areas)	English
Fluidex	automatic	English	none	English
Pascal	manual	French/English/Spanish	CT=(large areas), CC=?	French
CAB	manual	English	CT=(large areas), CC=?	English
Water Resources Abstracts	manual	English	CT= (DE=)	English
Waternet	manual	English	CT= (DE=)	English

2.3.2 "Water-related" databases and Mega-databases.

As water is an inherent part of the real world, the scope of water-related subjects could be widened in several directions. Information on "water" can be found either in large multi-disciplinary mega-databases like Compendex, NTIS, Pascal, or Chemical Abstracts (containing over 10 million abstracts), which are available on many hosts, or in specialised "water-related" databases on the environment (Environmental Bibliography, Enviroline, Pollution Abstracts, etc.), on biology (Biosis), on geology (Geoarchive, Geoline, Georef, etc.), on oceanography (Oceanic), on agriculture (AGRIS, Agricola, CAB, etc.) and so on, which are available sometimes on one host only.

2.3.3 Some remarks

It could be concluded that information on water-related subjects, for instance hydrology, or transport, or irrigation, is scattered over a large number of databases and information sources.

When one takes into consideration the fact that there are a number of abstract journals, which have no on-line equivalent, like Irricab produced in Israel, or International Civil Engineering Abstracts produced in Ireland, or the (West) German Dokumentation Wasser, the picture becomes even more complicated.

The development of databases stored on "new media" like the CD-ROM, as mentioned earlier, might be a solution for availability. Although the medium is rather well standardized, the retrieval software differs again for each product or supplier. In some cases the command language is similar to that of one of the existing hosts, which still makes experience necessary, but in most cases the retrieval systems are fully menu-driven.

Crucial is the fact that all databases mentioned are produced in the "developed" world using in majority sources produced in the same world. A lot of unnecesary duplication is present between all the databases, but information from developing countries is generally lacking.

Besides, a very important question is the applicability of the technology described in these databases for use in developing countries, but this is outside the scope of this paper.

According to Gotschalk (1984) of all internationally available databases at that time, less than 1% was produced in developing countries.

(No recent data are available. From the UNESCO Yearbook however it can be seen that the production of books in developing countries is steadily obtaining a bigger share of the total world production, but periodicals ? and reports??)

Exceptions are for instance the database Asian Geo at ESA/IRS, produced in Thailand, the database TROPAG (Tropical Agriculture) and IRC.DOC, the database of the International Reference Centre (IRC) in The Hague which is dealing with appropriate technology in the field of water supply and sanitation. This last database is loaded in-house with Minisis on an HP computer.

Solutions could be found in stimulating developing countries to either enter their material in an existing database or build databases themselves in some way. This has been one of the main fields of attention of the present working group, although no real progress has been achieved.

The cooperation between Delft Hydro and YUWAT has been set to provide a pilot project to exchange material in a machine-readable form, to serve as a model for developing countries. WRC, the producer of Aqualine is investigating this for the product of ENSIC from Bangkok. The IRC is also investigating this possibility, for instance the IRC.DOC database will be offered on CD-ROM together with the Repidisca database, which is produced in Colombia.

2.3.4 Some comparisons

Each database has its unique literature sources, but still there is an overlap in coverage between all of them. For instance, between Delft Hydro and FLUIDEX is a considerable overlap, attempts of the producers to cooperate to reduce the overlap failed however. There is also an overlap between Water Resources Abstracts, Waterlit, Waternet, Aqualine and Delft Hydro, especially when one considers the abstracted/scanned periodicals. A cooperation between the last two database producers has been established during 1987, to reduce the overlap and exchange abstracts.[1])

To compare some details of a number of databases, a very simple search has been executed: irrigat? and Senegal and au=hargreaves?. But this simple question already makes clear some of the issues mentioned earlier, like the large overlap in covering relatively easy-obtainable material and the lack of "grey" or report material. A weak spot in most databases appears to be the indexing concerning geographical areas in detail. This is of course a strong argument for using geologically oriented databases, which have generally a number of ways to search for a detailed location.
The results of this comparison have been summarized elsewhere.[ref 2]

---

[1] Some 700 abstracts from Delft Hydro have been entered into Aqualine. This agreement, however, lost its value after the discontinuation of Delft Hydro.

References to literature in which a more detailed comparison of "water" databases is made are cited in the list of references [3,4,5]. For instance Collis (1985) compared Aqualine, Water Resources Abstracts and Waterlit, Provost (1990) compared Aqualine, KIT Abstracts, Repindex, IRC.DOC and ENSIC.

A very global search in a number of databases on hydrology in the keywords gives the following results:

Database	size	Occurrences	Database	size	Ocurrences
NASA	1973390	3604	Inspec	4711863	2920
Chemabs	11462207	41	Pascal	5231296	5727
Compendex+	3264282	9725	Oceanic	261667	537
NTIS	1752465	11828	Fluidex	278164	304
			MatSci	1058326	895

Results of a global search on wastewater(w)treatment and waste(w)water(w)treatment.

Results of the first word ((w) means no words in between) are in column 2-4, the second word in columns 5-7. BI means occurrences in Basic Index, CT means keywords (controlled terms) and TI means occurrences in Title field.

Database	BI	CT	TI	BI	CT	TI
Oceanic	145	105	14	20	0	3
Pollution	10796	10008	1314	1499	1109	203
Fluidex	497	18	181	238	21	59
Aqualine	4068	3864	1149	1905	153	1041
ASFA	958	741	117	110	11	19
EAUDOC	401	0	401	40	0	40
CAB	734	233	314	1570	1319	82
NATO-PCO	57	0	15	8	0	2
Geobase	448	198	154	96	23	24
SWRA	14576	12246	1664	17137	16675	474

In some cases conclusions could be clear, EAUDOC uses French keywords, so English words only appear in the titles. In SWRA, Selected Water Resources Research the result is a little strange, and cannot be explained from the differences between US and UK English, like for instance in the case of CAB or Aqualine. For final conclusions additional tests are needed. Results of these test are not included here, but will be presented in the oral presentation.

3. References

[1] Gottschalk, C. (1984)
   Database management concerns at international level.
   In: "Database Management in Science and Technology, A CODATA
   Sourcebook on the Use of Computers in Data Activities," J.R. Rumble, Jr.
   and V.E. Hampel(ed.),
   Amsterdam, North-Holland, 1984, Chapter 10, pp. 241-263, 92 refs.

[2] de Mes, W.W. (1991)
   Oceans of information.... but where is the water?
   Paper 3rd International Seminar on the Management of Information related
   to Water and the Environment, Brussels, 14-15 November 1991, 20 p.

Database comparisons:

[3] Collis, R. (1985)
   A comparison of water-related online databases: Aqualine, Water Resources
   Abstracts, & Waterlit.
   London, City University, Department of Information Science, 1985,
   October, M. Sc. Thesis, approx. 100 pp.

[4] de Jong-Hofman, M.W.; Siebers, H.H. (1984)
   Experiences with online literature searching in a water-related subject field:
   Aqualine, Biosis, CA search and Pascal, compared using the
   ESA/information Retrieval System
   Online Review, 8(1984)1, Feb. p. 59-73

[5] Provost, F. (1990)
   Water related information on a global scale.
   Antwerpen, Universitaire Instelling, 1990, Thesis, 80 pp.

Working Group Papers:

[6] Nieuwenhuysen, P., Provost, F., de Mes, W., and Sicevic, M.
   Scientific and technical water-related documentary information in the
   UNESCO International Hydrological Programme (UNESCO-IHP Phase II,
   Project 17.1)
   Paris, Unesco, 1989, Technical Documents in Hydrology, SC-89/WS-49,
   51 p.

[7] Nieuwenhuysen, P., Provost, F., and de Mes, W.W.
   Information related to water and the environment: databases available online
   and on CD-ROM.
   Paris, Unesco, 1992, Technical Documents in Hydrology, IHP-IV Projects
   M-2-1 and M-2-2, 133 p.

112

Klaus-Dirk SCHMITZ
Fachhochschule Köln, Cologne, Germany

Multimedia Project and Glossary
on Wind Energy Terminology

Abstract: A team of linguists, computer scientists and experts of interactive technologies from Cologne, London and Naples is cooperating on a multimedia project to develop an interactive self-access learning programme for English as a foreign language for engineers. The project is essentially being funded by Fachhochschule Köln and the LINGUA programme of the European Union. The UK-based Wind Energy Group Ltd. is contributing the "real-world" background of the courseware. Running on a Multimedia PC with CD-ROM drive, soundcard, speakers and microphone under Windows 3.1, the "WINDS OF CHANGE" software allows intermediate to advanced level speakers of English as a foreign language to take part in active learning situations based on wind turbine technology, to learn effective strategies in English on a self-access basis, to record their own voice and compare it with model answers, to practice their English using language observed from engineers at work, and to consult support materials with a grammar and a glossary. The terminological glossary implemented as an on-line help facility contains about 1000 concept-oriented entries of wind energy and related engineering terms. Each entry is composed of the English term with its synonyms, the German equivalents, an English definition, an English example and sometimes a usage note. If a definition uses a term, that is defined elsewhere in the glossary as a seperate entry, a hyperlink mechanism allows the user to jump to this entry. The paper describes the basic features of the "WINDS OF CHANGE" courseware and the different approaches to implement a terminological on-line glossary in a CD-ROM and Windows-based environment.

1. Introduction

Since 1990, the Fachhochschule Köln is engaged in the development of interactive multi-media-based learning programmes forEnglish as a foreign language for engineers. In cooperation with South Bank University (London), Interactive Technologies (United Kingdom) and International Consultancy Services (Italy), the project "English for Engineers" started in 1992. This project is essentially being funded by the federal state of Nordrhein-Westfalen and the LINGUA programme of the European Union.

2. The "Winds of Change" programme

As the first step of the project work, a needs analysis was carried out among students of the Fachhochschule Köln and among practising engineers in Germany and Italy. The detailed analysis of interviews has indicated a great demand for skills and communication strategies in the field of language for general purpose as wells as for special purpose, and for oral competence in authentic situations.

On the basis of the evaluation of the needs analysis two main educational objectives of the programme were defined: the acquisition and/or practising of strategies for overcoming the communication problems as they occur in the professional every day life of engineers, and the brushing up of listening and reading comprehension as well as of verbal and writing skills.

The integration and realisation of these objectives in a multimedia environment with voice recording and play-back instead of developing a traditional course book aims at
- an increased motivation of the learner
- a high degree of interactivity
- an individual definition of learning speed and learning intensity
- the possibility for the learner to select individual learning strategies
- a fast and selective access to learning aids
- a direct evaluation of individual learning performance

The project partners agreed on choosing the subject field of wind energy as application area for the learning programme, since several engineering disciplines are involved in planing, constructing and operating a wind power station. The authentic background was given by an active collaboration of The Wind Energy Group Ltd., a UK based producer of wind turbines that generate electricity.

The first two experiments of implementing a multimedia learning programme for English for Engineers followed an analog/digital approach on the basis of a video disc and a CD-I (Compact Disc Interactive) environment approach in cooperation with Philips. But with the fast technological progress of IBM-compatible personal computers in speed, disk and memory capacity, and with the availability of multimedia components and authoring software for these types of computers, a more adequate and widespread hardware and software platform was found.

Developed with Authorware Professional and various tools for graphic, video and audio processing, the first Compact Disc (CD) with the title "THE WINDS OF CHANGE" has been produced. For using the CD, the learner needs a Multimedia PC (386 DX processor or better) with CD-ROM drive, VGA graphics adapter, SoundBlaster voice card with microphone and speakers (or headphones), and MS-Windows 3.1 or higher.

The main part of the learning programme consists of four modules
- "Overcoming Opposition"
- "Ensuring against Noise"
- "Unforeseen Difficulties"
- "Final Commissioning"
in which authentic *situations* in the whole process of setting up a wind power station are discussed between two English speaking partners. Each of these four modules is subdivided in two main situations with up to seven sections each.

The user can navigate through the sections via buttons, icons and hotspots. He can select the English and/or German display of the sentences on the screen and activate or deactivate the voice output. It is possible for the user to participate in the situations by taking over the role of one of the speakers, recording his own voice and comparing his voice with the authentic voice.

In addition to the situations, so called "strategic and structural *activities*" are implemented in the "Winds of Change" programme. In the strategic activities part the learner can practise and repeat specific communication strategies that are similar to the situations in the four main modules. The structural activities consist of exercises for deepening typical phenomena of the English grammar.

The third part of the courseware, the *resources*, are implemented as on-line help functions and are composed of:
- copyright information
- information on content and objectives of the courseware
- background information about The Wind Energy Group Ltd.
- programme orientation ("Where am I ?")
- user help
- situation oriented grammar
- terminological glossary

All three parts of the "Winds of Change" multimedia-based courseware will allow intermediate to advanced level speakers of English as a foreign language to
- take part in active learning situations on wind turbine technology
- learn effective strategies in English on a self-access basis
- record their own voice and compare it with model answers
- practice their English using language observed from engineers at work
- consult support materials including grammar and glossary.

3. The terminological glossary

3.1 Compilation of the contents

The first step in building up the terminological glossary for the "Winds of Change" courseware was to define the scope of subject fields that should be covered. The project team decided to include of course all wind energy related terms. These terms were supplemented by general engineering terminology necessary for understanding wind turbines and wind power stations. Economic efficiency and environmental protection related terms were occasionally added.

On the basis of terminological diploma theses and seminar papers, worked out by students of the Fachhochschule Köln, the terminology of the defined scope was collected and checked by the project team. For each concept the English and German terms (synonyms included), an English definition and context example, and grammatical and usage notes were recorded in a word processing file. For reasons of more convenient handling, the word processor's ASCII file was automatically transferred to the terminology management software MultiTerm 2 and afterwards to MultiTerm for Windows. In this way, a glossary of about 1000 concepts with almost 1300 English and 2200 German terms was compiled.

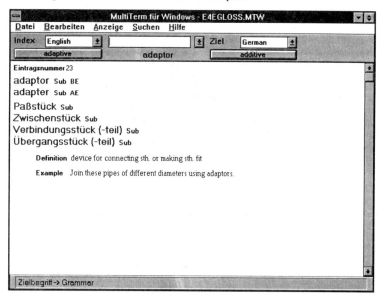

(Fig. 1: MultiTerm for Windows version of the glossary - entry screen)

3.2 Technical requirements for implementation

Since the terminological glossary should be used within the CD-ROM based "Winds of Change" software the implementation of the glossary has to meet some more or less technical requirements.

The software platform for the implementation of the glossary has to be compatible with Microsoft Windows 3.1 environment in order to allow the user to look up terms in the glossary while working with the courseware (on-line help). For the sake of user-friendliness the screen design and the user interface should follow the same design criteria as the courseware itself.

The educational objectives of the learning programme claim that the glossary software has to be able to present the terminological entries in two different ways. For the basic and first access to the glossary only the English terms, the definition and the context example should be visible. If the learner has problems to understand the meaning of the concept, he should be able to let the glossary software display the German equivalents.

The CD-ROM as a medium for distributing the courseware does not allow a writing access to the glossary (read only memory). Therefore the glossary software has not to deal with adding new entries, updating entries or deleting entries.

The best and most efficient way of implementing terminological glossaries is to adopt and adapt existing terminology management software such as MultiTerm for Windows. But the intended marketing of the courseware for commercial use and free dissemination to university institutes are in opposition to the licence agreements and royalties of the developer and/or local dealer of the terminology management software. Therefore, an implementation platform for the glossary had to be found, that uses licence-free run-time versions and that so allows marketing and spreading free of royalties.

3.3 Implementation of the glossary

For the first attempt the Microsoft Access Distribution Kit was chosen as development platform. The software is based on the Microsoft Access database management programme and meets the technical requirements as discussed in 3.2 (Windows, run-time version). After the definition of the database's entry structure, the terminological entries of the MultiTerm version of the glossary had to be transferred. Although the export routine of MultiTerm is very powerful and several additional macro routines were programmed, each terminological entry had to be individually modified and controlled, since the free entry structure of MultiTerm cannot be transformed to the fixed entry structure of Access by a completely automatic procedure. After all entries had been imported into the Access version of the glossary, the user interface, the screen masks and the handling and look-up menus were programmed in Access Basic (Visual Basic).

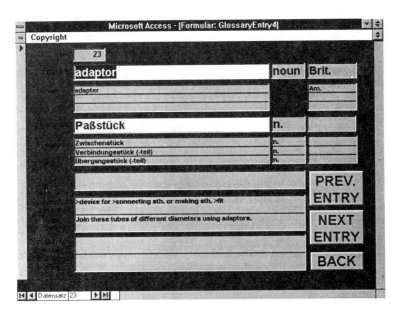

(Fig. 2: Access version of the glossary - entry screen)

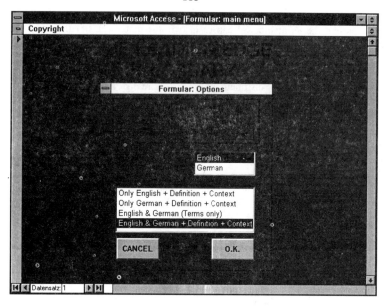

(Fig. 3: Access version of the Glossary - options screen)

When implementing the Access version of the terminological glossary, two main disadvantages of this approach arose. First of all, the Access database philosophy is not very adequate for handling concept-oriented terminology. The differing number of terms and synonyms per entry (from two to ten) and the necessary uniform access from all these terms to the concept entry are very difficult to implement. In addition to this, the intended references (links) from a term used in a definition to the related concept could not be realized. The second disadvantage of the Access solution is correlated with the screen design and the user interface. Although the possibilities of designing screen masks and menus are not very bad in Access Basic, they cannot reach the powerful potential of Authorware Professional, that is used as the authoring tool for the "Winds of Change" courseware. For this reason, the learner is confronted with three different screen designs and user interfaces: for Windows, for the learning programme and for the Access glossary.

To eliminate the discussed disadvantages, a second and final approach of implementing a glossary was selected which consists in modifying the Windows Help System for the purpose of managing terminology. The Windows Help System meets the technical requirements (Windows, run-time version) in an ideal way, and the screen design and the user interface follow a widespread common standard.

Since the Windows Help System is essentially based on text-oriented screens, it was an easy procedure to import the free-structured terminological entries with all terms and synonyms from the MultiTerm glossary version. Additional macro routines were programmed to optimize the lay-out of the entry, to implement the two different screen masks, and to generate the index terms for the look-up procedure. The learner can now look-up terms by using the traditional index search facility of the Windows Help System. The references from a term in a definition to a related concept could be realized by using the hyper-link features of the Help System. If desired, graphics can be integrated into the terminological entry in order to complement the definition as the concept explanation.

(Fig. 4: Windows Help System version of the glossary - entry screen)

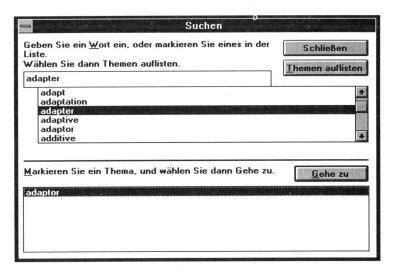

(Fig. 5: Windows Help System version of the glossary - look-up screen)

With the terminological glossary based on the Windows Help System, an adequate and efficient on-line help tool is available to support the user of the "Winds of Change" multimedia learning programme.

4. Future developments

The final version of the "Winds of Change" courseware will be available at the end of the year. Contacts with well-known publishing houses for marketing the courseware are have been established. The CD-ROM with the courseware is planned to be used in the polytechnic universities of the federal state of Nordrhein-Westfalen in cooperation with the Institut für Verbundstudiengänge der Fachhochschulen Nordrhein-Westfalens.

The implementation of the "Winds of Change" software has given quite a lot of experience in developing multimedia CD-ROM based courseware to the project team in Cologne and London. This experience should be transferred and applied to further courseware projects. Therefore, the "Winds of Change" programme is the first issue of a CD-ROM series called "The Leading Edge".

References

(1) Dette, K., Haupt, D., Polze, C. (Hrsg.): Multimedia und Computeranwendungen in der Lehre. Berlin, Heidelberg: Springer-Verlag 1992

(2) Ellis, G.: Learning to Learn English, A Course in Learner Training. Cambridge: Cambridge University Press 1989

(3) Graf, J. & Treplin, D.: Multimedia, Das Handbuch für interaktive Medien. Ulm: Neue Mediengesellschaft 1993

(4) Hill, D. A.: Visual Impact, Creative Language Learning through Pictures. Essex: Pilgrims Longman Resource Books 1990

(5) Hutchinson, T., Waters, A.: English for Specific Purpose, A Learning-centered Approach. Cambridge: Cambridge University Press 1989

(6) Kerridge, D.: Presenting Facts and Figures, Longman Business English Skills. Burnt Mill: Longman 1988

(7) Matthews C., Marino, J.: Professional Interactions, Oral Communication Skills in Science, Technology, and Medicine. N.Y. Englewood Cliffs: Prentice Hall Regents 1990

(8) Nemetz Robinson, G. L.: Crosscultural Understanding, English Language Teaching. International: Prentice Hall Regents 1985

(9) Soars, J. & L.: Headway, Student's Book (advanced). Oxford: Oxford University Press 1989

(10) Schmitz, K.-D., Schröter, F: Zur Entwicklung von Selbstlernmaterialien für die Fachsprache "Technisches Englisch". In: Fachsprachen und Übersetzungstheorie, VAKKI-Symposum XIII. Universität Vaasa: Vaasa 1993

(11) Steinbrink, B.: Multimedia, Einstieg in eine neue Technologie. Haar bei München: Markt & Technik Verlag 1992

(12) Trimble, L.: English for Science and Technology. A Discourse Approach, Cambridge Language Teaching Library. Cambridge: Cambridge University Press 1985

(13) Yates, C.St.I., Fitzpatrick, A.: Technical English for Industry, Coursebook. Burnt Mill: Longman 1988

S. Katusčak, J. Pajtik and Beseda I.
Faculty of Ecology, Technical University, Zvolen, Slovakia
SDVU, State Forest Products Research Institute, Bratislava

Classification and Knowledge Organisation in the Area of Ecological Quality of Materials

Abstract: The complex of characteristics expressing the behaviour and properties of the material in the environment can be named the *ecological quality (EQ) of the material*, including *eco-balances (EB)* and *eco-properties(EP)*. A variant of the classification of EQ and EP of materials is being proposed here. The study and quantifying of these characteristics of materials have been done in the last decade, with attention placed mainly on *eco-balances*, which are the data on the consumption of energy and raw materials and pollution of air, water and soil. *Ecological (or environmental) properties of material* are understood as the quantities expressing quantitatively partial interactions of the particular **m**aterial of men with **o**ther living components (M/O) of the environment, including *hygienic* and *toxic* properties and *biocompatibility/dwelling ability*, *biophysical*, *sensory* and *recycling properties*. E*Q*, *eco-balances* and *ecological (or environmental) properties of material* do not describe the environment itself but *the materials* in relation to their living environment.

1. Introduction

The quantifying of the *ecological (environmental) quality (EQ)* and *eco-properties (EP)* of materials is a relatively new area of science. It is an interdisciplinary area between material engineering and medicine with a utilisation of methods and approaches from chemistry, physics, biology and mathematics.

We use in this work the term *ecological* rather than e*nvironmental* to emphasize the interactions between materials and **living** parts of the environment; though, we are aware of the fact that the prefix *eco* is often used in Europe for expressing environmentally friendly activities, for names of *eco*-firms and *eco*- products, while the prefix *eco* stays for a synonym for "green", or "of high environmental quality", or "healthy", or in German terminology for a description of relatively new branches (Ökologie in Bau, Ökobilanzen, ökologische Bewertung, Ökoprofil, ökologisch Bauen mit Materialien, etc.).

Today, companies and consumers are as sensitive to the environment as they are sensitive to interest rates and inflation. The eco-sensitizing processes continue to enhance R&D, technology and quality of materials and products. To reach the goal of more eco-quality of materials and products in practice, e.g. in R&D and technology, it is necessary to replace the abstract goal *eco-quality* by a *quantified* set of measurable characteristics.

One of the first steps of scientific work on the quantifying of new phenomena is *classification*. The aim of this work is to propose a variant of the classification of EQ and EP of materials.

2. The Need for a Universal Language on Eco-quality and Biocompatibility of Materials

Not only scholars of pre-history agree that human progress was slow before language developed; only with language it became easier and faster to transmit the experience acquired by one person to other persons. It produces sufficient positive synergism for all human activities. The dependence of the development rate on the existence of language is generally valid for all human activities (1).

Until now no unambiguous basis existed upon which to build a universal language for the areas such as biocompatibility and dwelling ability of building materials. In the last few years, only the language of ecological balances (EB) has been sufficiently improved.The classification and knowledge organisation described below is proposed as one of the necessary steps in the creation of a universal language for ecological quality, biocompatibility and/or dwelling ability of materials.

3. The Role of Eco-properties in Technology, Testing and Comparisons of Materials

The present state of *quality testing* and *standardisation* is not advantageous for ecologically more progressive materials. The primary technical quantities developed mostly in the past centuries are still used for comparisons of materials and products. The processes of standartization have been initiated and influenced predominantly by the branches of high-tech, metal, inorganic and synthetic composite materials. Therefore the primarily *technical properties still play too strong a role* by comparison with materials and products. The weight of eco-properties in quality testing and standardization is very low and does not correspond to their importance. The ratio of the weight of eco-properties to purely technical properties should increase. For this purpose, however, it is necessary to *quantify and standartize* the characteristics of EQ in the form of a new, relatively independent group of material-engineering properties, which could be used by engineers, planners and other professionals, similar to the present primary mechanical, physical, chemical and other technical properties. Only in such a way will the technicians cease producing environmental problems that must be solved *ex post* under public pressure or because of environmental or health damages.

The present standard of *testing health effects* and related standards are not advantageous to ecologically more progressive materials, as long as mostly *toxic* effects, representing only a part of overall EQ, are measured.

Such comparison of materials results in a black and white approach: either the material is *negative (toxic)* or *OK (suitable, non-toxic, neutral)* according to present formal hygienic recommendations. If the materials are formally complying to present hygienic requirements, then there are no hygienic differences between them: for example hygienically OK building materials, such as plastics, iron-reinforced concrete, aluminium, wood or wool are all formally equal. False evaluation results if only toxicity is measured for example socks or other textile products made from wool would be of the same biocompatibility or EQ as products made from non-toxic polypropylene.

Building physics concentrates on more physical differences between materials; but here at least the problems of creating a standardized climate and thermal comfort from *any* material or panels have been solved. For example a building physicist, a specialist for wood materials, has to fulfil the thermotechnical requirements to the same degree as his colleague, a specialist e.g. from the area of the steel or concrete building materials. But again, if the standards (recommendations or requirements) on the thermal comfort are fulfilled, then the various materials - like wood, steel or concrete - are considered equal.

4. Classifications Related to the Eco-quality (EQ) of Materials

We have tried to create and classify a complex system of quantities to characterize ecological/ environmental quality (EQ) of materials, eco-balances and eco-properties. These quantities, similarly as measurable quantities generally, can serve better - in the development of new products, production, quality control, education and qualified advertisement of products - than general abstract terms alone.
Complex quality has the following groups of quality criteria:
technical(TQ)
economical(EcQ)
aesthetical(AQ)
ecological(EQ).

Classification of materials in relation to their EQ
  natural
          mineral
          biological (coming from plants or animals)
  synthetic/artificial
          identical with natural (e.g. ascorbic acid/vitamin C)
          similar to natural/easily biodegradable
          xenobiotic (the biodegradation is difficult)
  biomaterials (mostly highly biocompatible materials for medicine, e.g. implants)

There are 3 main *types of materials* from the point of view of their effects on health:
  biomaterials, or biomedical Materials (Mm)
          textiles, upholstery textiles or clothing textiles (Mt)
          materials for buildings and their interior (Mi).

Best elaborated is the biocompatibility in the first area of biomaterials. Here the contact between the biomedical material and living organism (Mm/O-contact) is the closest; the effect of the biomaterial on health is immediate. We consider it very important to analyse this area, especially the methodological approaches and methods of objectifying.

The second area of hygienic or physiological -hygienic properties and health effects of textiles is also relatively well elaborated. The upholstery and clothing textiles create a micro climate on the surface of the body. The rational for analyzing the knowledge and the methodology of objectifying Biocompatibility/Dwelling ability (BC/DA) and hygienic quality and effects of upholstery and clothing textiles on health is the same as for biomaterials. The Mt/O-contact and interactions between the textile materials (Mt) and body are not as immediate as those of biomaterials, but generally stronger than the contact with a building or interior materials.

The objectifying/measurement of the BC/DA-properties of the third group of the building/interior materials is the least developed area. The Mi/O-interactions are generally not so strong as Mm/O or Mt/O. On the other hand, these effects usually last longer, both if considered as the percentage of shorter periods of time (days, months) or years. The Mm/O-interactions are often very short, e.g. the contact of blood with the Mm of the extra corporal devices to the vascular system lasts only during the surgical intervention time, or in the case of Mm/O of the materials of drug delivery membranes used for controlled release of the pharmaceuticals it lasts only a fewe hours. The interactions between textiles and organism generally last for days or weeks; because after such a period the material is usually being washed or dry-cleaned and the foreign substances including toxic or bio-active ones are extracted through water or organic solvent solutions. The interactions between the building or interior materials and the organism, Mi/O, generally last for a great part of the day, for most days of a year, or for several generations.

Ecological quality (EQ) of materials. A complex of the categories, properties, and data expressing the *interactions* of the materials, with man and other living components of the environment, for the evaluation of the environmental dimensions of materials and products.

EQ Classification

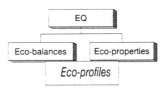

Fig. 1: The classification of ecological quality (EQ). The relations between eco- balances, -properties and -profiles.

The *EQ of materials* expresses the part of material quality related to the environment throughout all stages of its existence, from its origin, production, and use, to ageing and recycling.

The category of *ecological quality (EQ) of materials* was defined for example in 1988 at the conference: World-wide Research Strategies in Forest Products, at the 42nd Annual Meeting of the Forest Products Research Society in Quebec (2). The first attempt of classification was also made here. Simultaneously, the expectation was raised that the ecological quality of materials will be quantified and that eco-properties (EP) such as sensory properties, biocompatibility, eco-physical properties, recycling or some other properties of materials will be formulated gradually as a special, relatively independent group of properties in addition to the previously better-known and more commonly used groups of physical, mechanical, and chemical properties. During the past 6 years, substantial systematic work has been done on the quantifying of *ecological quality*, oriented mainly to *eco-balances*. Much less systematically, work has been done in the field which we call *eco-properties* of materials. Although it has been accepted that the questions of the *eco-quality* ("ökologische Qualität") of materials are urgent and relevant, the EQ is mainly understood as eco-balance (3).

Eco-balance is a set of data on the consumption of energy and raw-materials, air-, water-, and earth-pollution and effects on e.g. the CO_2-balance, throughout the whole life-cycle of a material/product (4,5,6).

Ecological properties (EP) of materials are the quantities descriptive of the partial interactions of the particular material with men and other living components (M/O-interactions) of the environment, including hygienic and toxicological properties and biocompatibility/dwelling ability, biophysical, sensory and recycling properties.

Eco-profile of material/product is a complex of quantified and non-quantified eco-balance data and eco-properties.

The quantified data should be preferred and as far as possible, related to one weight unit of the material (kg) for the sake of creating maximally compatible databases about various materials; only in interpretative later stages in various functional applications can the kg data be easily recalculated to the units of area (4) or volumes of products or to alternative products. When the basic kilogram eco-profiles are to be applied to flat products as paper, foils or textiles, which basic function is to cover some area, at otherwise comparable functional qualities, it is rational to recalculate the basic eco-profiles to area units. When comparing products - of otherwise comparable technical quality - (e.g. the four various milk packagings made from glass, aluminium, paper and plastics, or when comparing alternative windows made from various materials), the eco-profile must be related to the product itself. This is easily made from the kilogram fundamental profiles of particular materials according to their weight portion in the product (5). The early definitions of eco-profile came from the definition of ecology itself (4) as the science about the relationships of living organisms and the environment; but if one comes from this definition, one should not omit the most important effects on health or interactions with *man* and other organisms (7) during all stages of the material/product cycle (8) by concentrating only on pollution of air, water and soil through emissions from the production of a particular material or product. If one wants to base EP on the definition of ecology, one must consider, in the eco-profile of the material/product, the pollutions *and* the effects on man's health and other living organisms during utilisation.

In Table 1 we propose a variant of the classification of eco-properties.

The aim of this alternative classification is not to attain total accuracy, but to serve as a *creative impulse* and help in the search for new methods of quantifying said properties which have not yet been quantified.

It is suggested here to classify the EP into the four groups as follows:

- As the first group of EP we suggest *sensory properties* because the first aquaintance with a material and its EP comes through senses of interacting organisms (14).

- As the second group of EP we suggest *BC/DA* as the properties expressing results of material/ organism (M/O) interactions. They should express mainly the effects of materials on psychical or physical comfort or health through any methods or using any partial quantities.

- The third group are eco-physical and other primary properties of materials related to health effects.

- The fourth group of EP expresses the recycling ability of materials.

Sensory properties	Biocompatibility/ Dwelling ability (BC/DA)	Eco-physical and other primary properties	Recycling properties
Categories or quantities of materials or products detectable by sensors of human or other living organisms	Categories and quantities of materials or products expressing BC/DA through the anthropomorphous complexes of biological, sensory, psychometric, psychophysical or physical data/properties	Primary and other properties of materials or products hypothetically related to the eco-quality, biocompatibility /dwelling ability; for the sake of simplicity is the prefix eco- is not repeated in the following eco-related properties	Properties expressing the degradability (d.) and ability of materials or products to be recycled
anthropomorphous **optical** colour texture other o.p. (e.g. visual roughness)	**hygienic** toxic allergic toxic-allergic other h.p.	**thermal** t.conductivity t.comfort t.discomfort other t.p.	**biodegradability** through fungi in earth *in vivo* in higher organisms other bd.p.
tactile	**anthropocompatibility** /"**biocompatibility**" (e.g. of biomedical mat.)	optical	**ageing properties** photo-induced weathering
odor	**dwelling ability** (e.g. of materials/products in interiors)	electrical	**mechanical and chemical degradability**
aesthetic animal/plants compatibility	**structural**	re-use ability	
others	(degree of) **natural character**	chemical, physico-chemical and biochemical	**combustion/ pyrolytical ability, toxicity of the products**

Table 1 : Proposal of a classification of *ecological properties* (EP) of materials

Sensory properties of materials are secondary physical properties, which quantitatively express colour, texture, and other optical properties, as well as aesthetic, tactile, odour, acoustic and other properties which can be sensed, perceived and evaluated through the senses of living beings.

Living organisms are in permanent interactions with the material environment. They react to the external *stimuli* coming from the material environment. The quality and intensity of the stimuli acting of a material onto human senses depends on the *sensoric properties* of the material. The sensoric properties co-estimate the quality and intensity of the stimuli of the material environment on the human organism. These stimuli are either *under-threshold*, which the organism possibly perceives but does not react to or over-threshold to which the organism reacts.

The result of the stimuli of a material environment on an organism are *effects*. This class can be divided into:

- *physiological* (they do not disturb its internal environment and *homeostasis*, or the dynamic equilibrium of the relatively stable, internal environment; they do not damage the organism)
- *pathological* (they change sufficiently the stability of the internal environment; they damage the organism).

The reaction of living organism to the stimuli from a material environment are
- *automatic control* (of internal environment in a certain range without endangering the organism)
- *adaptation* (adjustment to the new conditions)
- *illness*.

The sensory properties of materials are generally known and described by (non-quantified) words, whereas technicians use for the description of a material some technical parameters semantically somewhat close to a sensory property in question (9,10,11,12). Gradually, new sensory properties better corresponding with human ways of perception and evaluation are being suggested, formulated and quantified.

Biocompatibility (BC) *with living organisms: man, animals and plants* generally expresses interactions of the material with living organisms. *Compatibility* means, non-technically speaking, the ability to agree, congruability, co-operation, harmony, good naturedness, combinability (Vertraeglichkeit, Vereinbarkeit); *compatible* means getting along with each other, going well together, complementing each other.

BC of materials is a category/property describing the behaviour of material in relation to the bio-system, human body or other living organism.

BC of biomedical materials: at the beginning and at the simplest level, this term was used to indicate a total absence of interaction between material and tissue (total inertness of material as the highest aim of its biocompatibility).
BC of biomedical materials has been redefined later (13) as the "ability of material to perform with an appropriate host (organism) response in specific application". After some modification, this definition can be used also for other classes of materials; more appropriate than inertness (see above) is the requirement, that the material and the organism/tissue should interact in the most appropriate way to ensure health. It is concerned with all aspects of the interaction that occur but focuses on the development of the response of the organism. This response of the organism, being the reaction of the organism to the material, controls the performance of the organism and can be subjective to some extent, but it depends also on the characteristics of the material and especially its sensory, physical, chemical, hygienic and other properties and on stability.

BC, as a quantity is neither *good* nor *bad*, just as temperature, time, dimension or other *primary* quantities are neither good nor bad. They are *relevant* or *irrelevant* to the phenomena in question and they achieve particular values for a specific function. Examples: temperature is a useful and *relevant* quantity for expressing the state in a refrigerator or on the sun, but nobody can say that a higher temperature is better than a lower one; it is quite good to have -10 °C in the freezer but a catastrophe to have -10 °C on the sun. But in both cases the temperature is relevant for the measurement, objectifying and improving the discourse about the performance of both systems.
The quantifying of the BC of materials is a relatively recent development. It is an interdisciplinary area between material engineering and medicine with a utilization of methods and approaches from chemistry, physics, biology, mathematics, etc.(14).

Biocompatibility and Dwelling Ability. Biocompatibility (BC) of material expresses the direct and indirect *interactions* of the material with living organisms. These interactions can arise either by *direct contact* of the living body/ tissue of the organism with the material or *indirectly through air, mass or energetic fields*. For the special case of interactions of man with materials used *in the dwelling environment* (construction and siding materials, floors, beds, chairs and other parts of furniture, etc.), the term *Dwelling Ability* (DA) has sometimes been used. We consider the BC to have a more general meaning than the DA: BC concerns all materials in contact with living organisms, such as biomedical materials, upholstery and clothing textiles, plastics, building materials, while DA concerns the materials in dwelling environment particularly.

The relation between BC and DA seems to be similar to that between e.g. mechanical properties - and strength, or between optical properties - and colour. Semantically *biocompatibility of a material* evokes a clear image of certain effects and a *certain* measure of agreeability between material and biological object. But it does not contain information on how far or how close the two objects are to each other; the word *compatibility* does *not* contain distance between the considered objects. It does not carry any information about the *direction* of the effects, about the *duration* of the interactions, whether these interactions are positive, zero (neutral) or negative ones. Therefore we think that biocompatibility of materials can be used as the general category for expressing any interactions between material and biological objects. And dwelling ability of dwelling materials is a term subordinated to biocompatibility. The term biocompatibility can be found in the literature until now, specifically in the medical area, and it's use there agrees with the general meaning of biocompatibility; the appropriate term for medical applications is biocompatibility of biomaterials or biocompatibility of biomedical materials.

From the methodological, experimental and instrumental point of view, it is rational to analyse *biocompatibility* and *dwelling ability* of materials in their mutual connection. The main reason is that the knowledge regarding the quantifying of *biocompatibility* is richer than the knowledge about *dwelling ability* of materials. Therefore the BC of other groups of materials, e.g. of biomedical or textile materials, can serve as the source of methodical and experimental approaches for the DA of building and other interior materials and products. In the following text, we use the symbol BC/DA for the biocompatibility and/or dwelling ability of materials in the environment.

The BC and DA are both *secondary* properties (15,16,17) and a part of the complex ecological quality of materials.

Hygienic and toxicological properties of materials express content and/or emission of harmful substances (such as formaldehyde, organic solvents, dioxin, vinyl chloride, chlorine containing low-molecular and macromolecular substances), expressed e.g. in micrograms per weight or volume unit of the material, or emission rate (ppm, mg/m^3). These characteristics express *negative* effects/properties of materials and products. In Table 1, the hygienic properties are understood as a part of the complex biocompatibility and dwelling ability of material. In comparison with the other groups of eco-properties/BC/DA (Table 1), this is already the best quantified group of properties.

The literature about the hygienic properties of wood, wood products and other building materials, about the allergic, toxic and allergic-toxic effects of individual harmful substances is rich, containing dynamically developed systems of data, standards, recommendations, laws and regulations, on national and international levels (see literature reviews about hygienic properties of building materials in e.g. (8), toxic and allergic properties (18), or of the gases from liquification of some treated products (19)).

Recycling properties express the degradability and ability of materials or products to be re-used and recycled. From the point of view of the methods of quantifying and measurement the most important are: *degradation*/stability (bio-, photo-, weather-, chemically- and mechanically- induced/performed degradation) and related kinetic parameters, re-use abilities and toxicity of degradation products, especially from pyrolysis or combustion.

5. Conclusions

The complex of characteristics expressing the behaviour and properties of the material in the environment can be named the *ecological quality (EQ) of the material*, including *eco-balances* and *-properties*.

The study and quantification of these characteristics of materials have been done in the past decade, with attention placed mainly on *eco-balances*, which are the data on the consumption of energy and raw materials and pollution of air, water and soil.

A variant of the classification of EQ and EP of materials has been proposed in this paper.

Properties of materials that express interactions between material and man or other living components of the environment play an important part. We call these properties *ecological (or environmental) properties of materials*. The *EQ of materials* expresses the part of material quality related to the environment throughout all stages of its existence, from its origin, production, and use, to ageing and recycling. The EQ does not describe the environment itself; it describes *the materials* in relation to their living environment.

References

(1) MacAdam, D.L.: Colour Measurement. Berlin, Heidelberg, New York: Springer-Verlag 1981.
(2) Katusčak, S.: Forest products research strategies in an era of ecological concern. Conference: Worldwide Research Strategies in Forest Product. FPRS 42nd Annual Meeting, Québec City, Hilton International, June 19-23 1988.
(3) Ökobilanzen und Recyclingfragen. Holz -Forschung und -Verwertung 44, (1992) 6, No.97.
(4) Hänger, M.: Basisdaten zur Ökobilanz von Holz. 20. Fortbildunskurs der SAH, 9./10. November 1988, SAH und LIGNUM Zürich 1988.
(5) Habersatter, K. and Widmer, F.: Ökobilanz von Packstoffen. Stand 1990. Bern: Bundesamt für Umwelt, Wald und Landschaft (BUWAL) 1991.
(6) Meier, K. Widmer, H.: Ökobilanzen - Grundlage füs umweltgerechtes Bauen. Unterlagen zum Seminar: Holz als Bau- und Brennstoff. Eine ökologische Bewertung. Bern:Bundesamt für Konjukturfragen Januar 1991.
(7) Ökoprofil von Holz. Untersuchungen zur Ökobilanz von Holz als Baustoff. Hg. Bern: Bundesamt für Konjukturfragen, Schriftenreihe IP Holz 1990.
(8) Schwarz, J.: Ökologie im Bau. Entscheidungen zur Beurteilung und Auswahl von Baumaterialien. Bern, Stuttgart: Verlag Paul Haupt 1991.
(9) Kollmann, F.: Technologie des Holzes. Berlin: Verlag von Julius Springer 1936.
(10) Bosshard, H. H.: Holzkunde. Basel, Boston, Stuttgart: Birkhäuser Verlag 1984.
(11) Kaufmann, H.: Eigenschaften des Schweizer Holzes: Asthetische Untersuchungen. Schlussbericht NFP-12-Projekt 4.052-0.87.12 (Fichtenholz) des Schweizerischen Nationalfonds 1988.
(12) Hoadley, B.R.: Identifying Wood. Newtown U.S.A.: The Tauton Press 1990.
(13) Williams, D.F.(ed.): Definitions in Biomaterials. Amsterdam: Elsevier 1987. p.6-7.
(14) Katusčak, S. and Gfeller B.: Study on Biocompatibility and Dwelling Ability of Wood in comparison with other materials. Res. Rep.57/93, Bratislava: SDVÚ 1993.
(15) Galileo Galilei 1623 : Il Saggitore. The translation by A.C. Danto: *Introduction to Contemporary Civilization in the West.* 2nd ed.; New York, Columbia University Press, 1954.
(16) Russel, B. 1910: The relation of sense-data to physics. Scientia, 1910, republished in: Mysticism and Logic. George Allen & Unwin, Ltd., London, 1917.
(17) Massof, R.W., and Bird, J.F. : A general zone theory of color and brightness vision. I. Basic formulation. J.Opt.Soc.Am. 68, (1978) No.11. p. 1465-1471.
(18) Hausen, B.H. 1981: Woods Injurios to Human Health. Berlin, New York: Walter de Gruyter.
(19) Hirata, T. et al. : Combustion gas toxicity, hygroscopicity, and adhesive strength of plywood treated with flame retardants. Wood Sci. Technol. 26, (1992) p.461-473.

Krystyna Siwek
Information Processing Centre, Warsaw, Poland

Linguistic Tools for Knowledge Presentation in Bilingual (Polish/English) Databases on Research and Development

Abstract: The Information Processing Centre (OPI) manages databases on R and D. For an international co-operation, English versions are prepared and CD-ROM and online services are being performed. The transformation process leads to various questions concerning the lack of standardization and the inefficiency of actually applied linguistic tools. Short characteristics of OPI databases - record contents, classification schemes and indexing methods are presented. The National Citation Report for Poland from ISI (Philadelphia, USA) is being compared with the Polish R and D Directory and the Who is Who in Polish Science for scientometric and linguistic analyses - with an interdisciplinary and a discipline-orientation. Participation in the Eureka program in the environmental sciences, in agriculture and biotechnology are mentioned as examples of international cooperation. As an example of efforts of other information services in the R and D area the project of a database of Polish journals abstracts is described.

1. Introduction

Bibliographical and factual databases designed and managed in information centres of The Polish Academy of Sciences (Polska Akademia Nauk - PAN) and The State Committee for Scientific Research (Komitet Badan Naukowych - KBN) deal with the entire spectrum of Polish science. Most of them are in Polish, but lately English versions were prepared for the purposes of international co-operation, information interchange and Polish research and development promotion. Some of the newly designed databases are in English only and some of them have been both, Polish and English elements.

"The translator's handbook" (1) deals thoroughly with various translation problems. Its scope is broad, it will not be discussed here. It should perhaps be taken into account that for many years the obligatory foreign language at the primary and secondary schools, as in any other postcommunist countries, has been Russian. As a result there is a lack of the middle level, qualified staff for information processing and editorial work in English.

Besides translation troubles there are many methodological tasks to be coped with as e.g. the problem of subjectivity and objectivity in classification and indexing, e.g. in the cases of the self-definition of researchers' specialization, of discrepancy between the name of an institution and the area of its real activity etc.

Standardization and unification on many levels - from database scope and comprehensiveness, bibliographic details, indexing languages to file organization and software have been the main themes of the report of the International Council for Scientific and Technical Information Group on Interdisciplinary Searching (2). The report concluded in the consideration that database producers should adopt national and international standards in all application cases.

Terminological incoherence in description of the same objects or phenomena in different databases or even in one database is the reason of inappropriate and misleading interpretation. Various linguistic tools such as vocabularies, dictionaries, classification and notation schemes, controlled keywords and thesauri - necessary in indexing and retrieving process - must be analyzed.

In the Information Processing Center (Osrodek Przetwarzania Informacji - OPI), which is The State Committee for Scientific Research, a launch team has been established recently for unification and standarization of OPI databases, because of the necessity arising of adoption on-line and CD-ROM versions. In the first stage of the databases, input and output information, indexing and retrieval tools were analyzed and a project of transformation will be designed and evaluated. In the meantime English translations will be made. One of initial efforts should concentrate on the unification of linguistic tools in the R and D area in Poland related to European and worldwide standards.

2. Characteristics of OPI databases

OPI manages databases on science policy such as:
- a Directory of the Polish Science Database (Polish and English version)
- a Who is Who in Polish Science
- a Dissertations Database
- a Research in Progress and Research Reports Database SYNABA
- a Reference Database to Information Resources

In the experimental stage are :
- a Database on Polish scientists and experts abroad
- a Database on research equipment named APARATURA

1) Directory of Polish Science
The Directory of Polish Science is managed as a computer database to be published biannually (3). The English automated version is already prepared and the hardcopy publication will be disseminated in September 1994 (4). The scope of the Directory is the organizational structure of Polish science. It consists of over 5000 items - R and D institutions. (The English version is narrower - without military and paramilitary R and D institutions). Each item contains an identification number, institutional name, address, telephones, telefax and email, name of the head, main activities (textual information), names of professors and assistant professors, number of doctors (PhD), postgraduate and doctoral studies, serial publications. Inverted indexes are provided for personal and corporate names.

2) Who is Who in Polish Science
The Who is Who in the Polish Science Database (in Polish only) corresponds to the Directory, because entries (over 50,000) concern scientists - professors, assistant professors and doctors - mentioned in the Directory. Entries contain names, titles, specialities, addresses and references to identification numbers of institutions in the Directory so the Who is Who database (and publication) is a sort of extended index to the Directory.

3) Dissertations Database
Additionaly there is a separate database named Dissertations (Doktoraty i Habilitacje) mainly for statistical purposes. Every record contains, besides personal and CV data, detailed information on doctoral and doctor habilitation degree.
Entries include the dissertation title, promotor name, reserarch institution where degree has been achieved, doctoral discipline and specialization - according to the official governmental terminology. A thorough classification scheme is used for all fields in records for the production of statistical tables.

4) Research in Progress and Research Reports Databasem SYNABA
Dissertations on the other hand are treated as a sort of reseach reports and for that reason they join the SYNABA database.

All the databases mentioned above have a relational dBase structure but SYNABA is implemented on CDS/ISIS microsoftware.
Till now SYNABA - the database on R and D projects and research reports in Poland (over 90 000 items) - is exploited in the Polish version only, but from the beginning of 1995 some additional elements in English will be included. The SYNABA record consists of the project type and title, authors and supervisor names, institutional affiliations - names and addresses (the institution where a project is conducted, the sponsoring institution, the institution where the final product is going to be utilized),

international co-operation (countries, international organizations and programmes names), financial data (especially KBN support), bibliographical information if the report has been published already, terms of beginning and end of a project, short textual characteristics and - finally - classification and indexes. It must be mentioned here, that SYNABA (with some modifications) has been conducted by the former Scientific, Technical and Economic Information Centre as an information retrieval system for 15 years and has been managed as a manual card-index system already since the early 70ies. At the very beginning, the Universal Decimal Classification was used to define the area of research reports. As a next item, the classification scheme developed in the International Scientific and Technical Information Centre in Moscow (USSR), translated and adapted (5) was used because SYNABA was destined to be the national subsystem of The International Information System on Research Reports (for communist countries). This classification system was adequate for Soviet science and economy and was partly incoherent for the Polish reality. (Gaps caused through the incompatibilities of culture in knowledge interpretation have been analyzed in (6).) Thereafter the Polish Subject Classification was created (7) and after various corrections it is now used for classifying SYNABA entries. The official territorial classification of enterprises is used for coding institutional affiliations. Besides this, the KBN Classification of disciplines and the statistical classification of economy have been implemented in the last two years. The detailed subject characterization of research projects is provided by a set of keywords. English elements which will be added in the nearest future are title and keywords only. It is apparently an unsatisfactory solution and can be approved only for a short period of time. Unfortunately the new project of the SYNABA questionnaire contains only these two English data. Preparation of a full English version with modified indexing tools should be taken into consideration. As it would affect a collaborative English translation, editors must decide on British or American version, otherwise there will be gaps in the alphabetically ordered sequences, indexes, etc. and this version should be common for every one of the OPI databases.

All these databases are valuable sources of nationwide information on achievements and gaps in Polish science - on institutional and personal ranking, in all domaines and in separate disciplines, what is a main point of interest for government agencies for science policy as e.g. the State Committee for Scientific Research. On the other hand, information services provided by OPI are highly evaluated by average users for their relatively satisfactory completeness and currentness. A problem to solve is incompatibility of files structure, terminology, lack of a common classification system and indexes. Therefore compiling records from different databases is very difficult. Terminological unification on institutional names and scientific discipline levels is also a most urgent task.

3. The National Citation Report for Poland

OPI databases provide a unique possibility of comparative nationwide studies (among others) on linguistic aspects of scientific information. An additional advantage is anticipated in correlation of OPI databases with the resources of the Institute of Sientific Information in Philadelphia (USA).

The National Citation Report (NCR) for Poland (8) - an extraction from the database in the Institute for Scientific Information (ISI) in Philadelphia provides data on stratification of Polish science in worldwide competition. Together with elements of Polish databases on R and D Institutions and the Who's Who in Polish Science a relational database will be established. At first some necessary corrections in NCR records must be done. For the control of authors names and affiliations they have to be identified in Polish resources.

The NCR includes about 71,000 papers of Polish authors indexed in ISI the database during the period 1981-92. Actualized editions of NCR will be provided as well. The updating of the file for 1993 will bei finished in the beginning of 1994 and so on.

For each paper information is provided on its title, the names of all authors and their addresses, journal title, volume, initial page number, year of publication, year of including the record in the ISI database and its citation counts - in total and year by year. There are some limitations incomparison with SciSearch or the SCI CD-ROM. The information on the list of references for every source article is not provided. Not included are some Polish authors who work at present in other countries when there is not explicite information about their permanent affiliation. The advantage of the NCR is the possibility of analyzing citation counts for all co-authors.

Some difficulties have been found with proper identification of authors and their affiliations. The most common problems NCR database users my deal with are:

- there are no Polish (and any others as well) diacritical marks
- there are only initials of Christian names. Poles in general do not use the second name so there is only one letter indicating the first name for identification purposes
- there is no information according sex
- there are no separators between two surnames - often used in Poland by married women e.g. instead of [Nowak-Kowalska Anna] the NCR database has [NOWAKKOWALSKA A]
- in the Polish databases on R and D only those names of scientists with a doctoral degree are included, so Polish authors included in the NCR without that degree will not be verified in these databases
- institutional names are translated in a different way in the NCR and Polish R and D Institutions database
- there are errors in dividing institutional name and address concerning six components: organization, department, laboratory, section, city, country and abbreviations (for reasons of keeping the record structure small) thus the data are not always clear
- during the past 10 years the organizational structure of Polish R and D institutions has been changed, thus information on authors' present affiliations may be misleading and the only way for a succesfull identification is to check the old editions of directories and any other information sources on science in Poland, because unfortunately a comprehensive databases containing archival information is not available concerning these data - even in the Polish language
- the Current Contents Category Codes and the Product Codes, items of the ISI classification are slightly different from the terminology of scientific disciplines as used in Poland. The Product Codes are titles of ISI Current Contents series (e.g. A - Current Contents /Agriculture, Biology and Environmental Science). Category Codes contain three letters as discipline abbreviations. The full notation consists of Category and Product Codes. For example:

    ECE T Environmental/Civil
    ENV A Environment/Ecology
    ENV C Environmental & Social Medicine
    SOC C Environmental & Social Medicine

Such codes do not describe a single paper but journal titles. Particular keywords can be obtained from paper titles for detailed linguistic analysis purposes.

4. Co-operation with Eureka

Besides these long range goals some mission oriented temporary actions are performed, e.g. the EUREKA Brokerage Event "R and D Co-operation with Poland" Warsaw - May 25/26 1994, organized by The State Committee for Scientific Research and the Information Processing Center from Polish side and by The Netherlands Eureka Office. Three main areas of interest of that meeting are: environment, energy and biotechnology. Six parallel sessions were held: 1. Food production, 2. Monitoring systems for the environment, 3. Waste water treatment, 4. Solid waste treatment, 5. Biotechnological processes, 6. Medical applications. One of the advantages should be a cooperation between the Eureka database on R and D projects and the English version of SYNABA. In the Eureka database the British Standards Institution Codes and the Standard Industrial Classification Codes are used. Of course file organization and textual content are different from SYNABA but there are many common elements, so information interchange seems real.

5. Polish Scientific Journals Contents Database

Many Polish scientific journals, especially those edited by institutes and learned societes of The Polish Academy of Sciences, are published in English. Others have English summaries, abstracts or even lists of contents. The Scientific Information Center of The Polish Academy of Sciences has developed the Polish Scientific Journals Contents database to be disseminated frequently on diskettes. This database is still in an experimental stage, a prototype software based on Micro CDS/ISIS has been performed and abstracts from 150 journals divided into three series (Humanities and Social Sciences, Life Sciences, Math.-Phys. and Technology) are included. Some problems in designing bibliographic interdisciplinary databases in English could be generalized and some are specific to the limitations of the Micro ISIS software. The necessity of leaving or dismissing national character sets was discussed too. For example CCoD - Current Contents on Diskettes database from ISI of Philadelphia does not use diacritics at all. The second question concerned mathematical, numerical, chemical etc. terms - in abstracts, titles and keywords. At last the Polish or English version of institutional names and addresses was under consideration and - finally - discipline classification. The necessity of authority files - mainly for corporate names, journal titles and disciplines was felt during the testing of search strategies and input procedures. So the same conclusion as in case OPI databases: to implement standard Polish/English authority files for the R and D area. Unfortunately because of the reorganisation of the Polish Academy of Sciences' information services activities concerning Polish Scientific Journals Contents stopped for the time being.

6. Conclusions

Fortunately, the theoretical and empirical linguistic and scientometric research program could be connected with the practical and necessary activities of improving information services in the R and D area. Standardization of terminology, indexing and bibliographic rules, data content presentation allow for searches in various databases. Since the scope of these databases is broad and satisfactorily comprehensiven, interdisciplinary searching, discipline oriented analyses and statistical aggregations can be performed.
The progress report dealing in greater detail with interdisciplinary and specialized linguistic tools together with scientometric analyses of databases on R and D will concern OPI databases and NCR for Poland's databases.
One of the aims that the Information Processing Center should achieve is a nationwide authority file of Polish scientific institutions. A reasonable solution would be cooperation with other organizations, e.g. - with The National Library and, perhaps with university libraries which have developed recently, for cataloguing purposes, but very useful in factographic databases, the authority file of the corporate names as a partof the programme of implementation the VTLS (9).
A contribution of The Ministry of National Education is "The Directory of Polish Universities and other Higher Educational Institutions" (10). The Business Foundation has published "Innovation, Research & Development (11) - similar to OPI's Directory but not so comprehensive.
At present Polish databases are in the transformation process for being fully accessible via international information networks, so there is a lot to do in coping with all the issues mentioned above for adopting services based on new technology. The cooperation with ASSISTANCE, the Central European Initiative and the European Union projects should be helpful.

References

(1) Picken C., (Ed.): The translator's handbook. London: Aslib 1989. 382p.

(2) Weisgerber D.W.: Interdisciplinary searching: problems and suggested remedies. A report from the ICSTI Group on Interdisciplinary Searching. J. Doc. 49(1993) No.3, p.231-254

(3) Informator Nauki Polskiej 1992/93. Wydanie XXV. Warszawa: Osrodek Przetwarzania Informacji, 1993.

(4) Directory of Polish Science 1994. Warszawa: Osrodek Przetwarzania Informacji, 1994

(5) Wykaz tematyczny dla klasyfikowania materialow informacyjnych (wraz z symbolami UKD), Warszawa, Centrum INTE, 1977

(6) Shreider Y.A.: The cognitive approach: a tool to resolve the opposition between technical and cultural aspects of knowledge. Int. Forum on Inform.& Doc. 17(1992) No. 2, p.3-6

(7) Polska Klasyfikacja Tematyczna, Warszawa, Instytut INTE, 1985

(8) National Citation Report. Inst. for Scientific Information, Philadelphia PA, 1993 10p. [brochure]

(9) Wozniak, J. (Ed): Kartoteka wzorcowa jezyka KABA. Warszawa: Wyd.SBP 1994. 162p.

(10) Directory of Polish Universities and other Higher Educational Institutions. Rep. of Poland, Ministry of National Education, Warszawa 1992.

(11) Business Foundation - Innovation, Research & Development, Warszawa 1993.

Irene Wormell
Royal School of Librarianship, Copenhagen, Denmark

SAP-Indexing for the Exploration of the Rich Topical Context of Books and for Accessing Smaller Semantic Entities

Abstract: With the SAP-indexing method new and innovative ideas for the development of modern subject representation forms and retrieval tools are put into practice, providing access to detailed and specific subject information. This approach eliminates the known shortcomings of existing bibliographic catalogues, in which the extensive contents of books find generally only a very poor representation. Using the SAP-indexing method, existing MARC-like bibliographic catalogue records were enriched with terms selected from the tables of contents and from back-of-the-book indexes with the exact location of page numbers relating to a concept treated in a book. This method was successfully used at Lund University Library, as well as in several other settings, for accessing specific parts of documents or in order to retrieve factual data presented in the form of tables and graphs scattered throughout a publication. It is suggested to consider smaller semantic entities than the entire physical document for representation and information retrieval in a sensible way.

1.   Introduction

In the middle of 1970s the issue of *environmental protection* and methods for *alternative energy production* was an expanding subject area in the Swedish research and academic libraries where the existing standard library classification systems did not offer   sufficient retrieval capabilities for this emerging multidisciplinary subject area. Topics dealing with complete scientific, social and political issues could not be "squeezed" into the categories of the rigid and traditional classification schemes. In 1977 The Swedish Research Council for Scientific and Technical Information awarded the author a fellowship to develop a new version of the library catalogue with extended subject access to library collections. She became affiliated with the Lund University Library and   started to design and establish a test-data base for augmented subject description for an online catalogue of books in the field of ecology and environment.

In the view of given bibliographic conditions in Sweden, it was clear that the new way of producing subject descriptions of books could not involve deeper analysis of the documents by the indexer. It would be an economically feasible way to improve subject access, and it would be based on existing and available technologies. The results would also provide a basis for suitable form for online interactions in heterogenous user environment. (It should be noticed that this project was designed long before the start of existing OPAC developments).

Meeting the demand of improved retrieval possibilities and more effective search aids for the user of online library catalogues in this subject area, it was suggested to use the inherent attributes of the documents, instead of the meta-term descriptions of subject contents in the form of controlled vocabularies or thesaurus terms, - added by librarians or others responsible for producing access to catalogued publications. The most appreciated feature of the SAP-file was that it permitted searching of not only a broad subject area but even a specific aspect of a book's treatment of that subject. The selected word strings and phrases all pointed to the significant topics discussed in a book, giving also the location where in the book to find them.

The underlying idea was to provide entries to those semantically rich but confined parts of the physical documents. As consequences of latest progress in information technology and business this idea is again in the focus of interest. The necessity to create multi-dimensional subject access to stored knowledge which, in a highly flexible way, may satisfy the variety of users and types of information requirements gives for SAP-indexing a new currency. Recent IR-developments suggest to view the smaller semantic entities than the physical documents as the proper entities for representation of information in a sensible way. The advantages are that access to the document in its physical sense is not lost, but enhanced, and the intellectual access to the variety of potential spots of users' interest users is multiplied.

2.   SAP indexing methodology

Subject Access Project (SAP) was aimed to test the possible usefulness to the content tables, back-of-the-book-indexes, captions of tables and diagrams for creating detailed subject description for the monographic publications in modern online catalogues. The suggestion that the *inherent* attributes of the documents can be used as the source for their representation, has been the underlying theoretical framework for the project - both in the USA at Syracuse University, School of Information Studies (Atherton 78) as well in Sweden at Lund University Library (Wormell 85).

The idea was to enhance subject access to monographic publications by augmenting MARC-like catalogue records with natural language terms selected from the table of contents and the back-of-the-book-indexes, and to provide exact location of page(s) where the topic is treated in the book. The selection of index terms was conducted by easy, quantitative rules. The augmentation of subject description was running up to an average of 300 words per title or about 30 entries per book. An entry in the catalogue record was a string of terms in a *contextual* form such as, a chapter heading, a subheading, or selected index terms from the back-of-the-book-index.

Figure 1. is giving a sample of the prepared input for a book. The selected subject headings are underlined and supplied with page ranges.

Figure 2. shows the corresponding SAP-record, consisting of the ISBN number, short bibliographic data, Content Terms (CT) and Index Terms (IT) grouped in separate fields.

Figure 3. shows a simple example of search output from the first Swedish SAP-file, containing 400 Swedish and English monographs in the field of ecology. The search result is presented in the shortest display format. Observe that "hits" are always represented in *contextual strings*. In today's systems there are available technologies for the exhibition of terms in a much more practical way, for instance to highlight the searched terms in string context, which enable the user to view his search terms and make quick relevance judgments.

3.   Access to factual data

In 1979 the SAP project at the Lund University Library advanced to its second phase aimed to enhance subject access to a very significant source of evaluated and scrutinized factual information in Sweden: the **SOU-series** (Statens Offentliga Utredningar = Swedish Government Official Reports), a series of reports containing a great amount of descriptive, factual and numerical data in the form of tables and graphs. The idea was to create direct access to the rich contents of non-textual materials too, captured in the form of tables and graphs - a facility which until now was not really given to the users of online catalogues.

It is an old tradition of Swedish parlamentarism that when new legislation is proposed, the Government establishes a committee to explore and analyze all facts of importance related to the proposal. This committee activity is recognized as an essential precondition for good public policy-making. As a rule, the committees submit extraordinarily comprehensive and detailed reports, which are often also of considerable scientific value. Prior to the drafting of a bill, the Government sends the report on it to various government bodies, agencies, organizations and other who may be concerned. They have to review the recommendations and may propose changes. In this way, all the facts are evaluated and scrutinized, and it would be realistic to say that the data constitute a reliable source of information.

100-130 reports appear annually in the SOU-series and they contain a great deal of factual and numerical data in the form of tables and graphs, acquired through the strenuous efforts of social scientists, civil servants, experts, research workers etc., and through the allocation of great financial resources. But because of the lack of efficient retrieval capabilities, the rich contents of these reports are not sufficiently explored and applied outside the realm of the committee. Here too, as in the case of monographs, the data are buried in the publication and are scattered throughout the text. Full usage of the data cannot be expected, because there are no direct access mechanisms, except browsing through hundreds of items.

Each report usually has a short title, sometimes ambiguously worded in order to attract interest, but not very helpful to the reader, e.g. a sloganlike, strained title such as "Ren tur" (SOU 1974:44), which has the double meaning of "Pure Chance" or Clean Cruise". Only through a subtitle, namely "A Programme for Environment-Safe Sea-Transports", does it become clear that the latter meaning is the correct one. From the subtitle (which is not always searchable in the catalogue), we can infer that the report contains data about the environment. It is, however, difficult to imagine from the titles that environmental data can also be found in reports with titles such as "Children's Accidents" or in a report the title of which would read in English "Disabled - Integrated - Normalized - Evaluated" (SOU 1980:34). If we were to rely upon title searches in cases as these, many of the existing retrieval mechanisms in todays' libraries would not be very helpful.

The quality and the reliability of numeric data are the crucial problems of factual retrieval. Therefore strenuous efforts have been made, e.g. by CODATA, to give international recommendations for labelling numerical data that may be cited independently; Figures, Tables and Graphs are to be considered as logically complete units, independent of the main text. If we presuppose that authors and editors are aware of these recommendations, we can believe that the captions of tables and graphs are reliable text strings that can be searched.

The SOU-database was covering a range of ten years, 1970-1981, containing approx. 1,200 reports. By the time SAP-indexing had been recognised as a promising new way to produce detailed subject descriptions, and the principles became a guideline for the development of specific in-house databases in several environments. As a convincing result of the research project, the SOU-database was taken over in 1983 by one of the large commercial hosts in Sweden (DAFA), and it has been continuously updated and used since. It is publicly available today as an integral part of the Swedish information system for legislation and public administration (RÄTTSDATA).

4.   Applications of the SAP-indexing methodology

To illustrate the different indexing approaches to representing the topic versus the contents of a monographic publication, one may refer to a specific application of SAP indexing by the Danish Pedagogical Library (Poulsen, 90). The SAP indexing methodology served as an easy and economic way to identify the specific contents of pedagogical publications. Thus, the intellectual effort of the indexer was concentrated on the identification of those scientific paradigms and political viewpoints which are related to the pedagogical subject matter treated in each publication. This type of information on viewpoints and scientific treatment of topics are normally not made explicit in the Table of Contents, but has to be added by a domain expert. The SAP methodology was hence used to represent document contents whereas the assigned indexing provided access to the "hidden" aspects of this contents. This combined approach of NLR and assigned indexing demonstrates the usefullness of complementary methods in enhanced subject indexing.

Another application of combining indexing methods in the area of OPAC's is reported by S.A. Cousins from Aberystwyth, UK (Cousins, 92). In her experimental databases, content bearing terms were extracted from the table of contents and back-of-the-book-indexes using the SAP-indexing technique. These selected natural language representations were then translated into PRECIS strings to make sure that the indexing language was adequately specific and standardized to meet the users' queries (subject queries were also collected and indexed by PRECIS). Searches were carried out on three test databases comparing search performances between 1) simple MARC records alone; 2) records enhanced with word strings selected by SAP; and 3) records in which SAP terms were structured according to PRECIS. With respect to the types of subject representation the test results not surprisingly demonstrate that both natural language representations (SAP) and the PRECIS enhanced records increased the recall compared to the simple MARC records. However, the difference between the performance of SAP and PRECIS records was so small that natural language can be regarded as good as controlled vocabulary for enhancing the subject contents of and access to OPACs. Also within the framework of the SAP applications in UK, the retrieval value of back-of-the-book-index terms has been tested against terms derived from the table of contents. This test showed that approx. 23% of the relevant documents were retrieved using terms from the back-of-the-book-index for subject representation. Thus, index terms derived from the table of contents seem to have greater retrieval potentials (Cousins, 92).

Additional experimental evidence gained from SAP suggests that captions of the tables and graphs are usually more self-explanatory than the some times fancy titles or chapter headings of the books. Thus, one must be aware of the fact that the more specific and precise terms required for retrieval of factual data are to be found in the captions.

The project at the Göteborg University Library, 1990-91, titled "Chapter indexing. Enchanging catalogue records with chapter titles, subject indexes and bibliographies" was aimed to test OCR technique in connection with enhancement of conventional catalogue records (Cavallin, 91).

Chapter headings of 932 monographs in the humanities and social sciences were read into a computer using OCR technique. Based on the results of 89 search questions running against the database, bibliographical citations and chapter headings were found to be the most efficient means of retrieving relevant references. After adjustments concerning size and number almost 30% of the relevant references turned out to have been retrieved through the chapter headings. It was also clear that the precision rate in the chapter headings results was quite good. The report, prepared for BIBSAM - Office for National Planning and Co-ordination, Stockholm, recommends enhancements of catalogue records with chapter headings. It is concluded that it is still up to the libraries to decide whether they are prepared to pay the extra money required (approx. £1 per book) to add the chapter headings, and to provide the appropriate equipment and the extra memory capacity necessary (3 times more than what is needed for a normal catalogue record).

Indexers and LIS professionals in Denmark working within environmental and human rights organizations also applied the SAP indexing technique for subject representation in their international information systems which are used globally. In the environment where there is a big difference or there is a lack of bibliographic standards, SAP indexing is functioning as facilitator of global communication: the easy and straightforeword way to use the inherent attributes of documents for subject representation is a practical basis to built up e.g., a clearinghouse that could refer enquiring groups to those with relevant expertise. Since this indexing technique does not require the qualified knowledge and experience of a professional indexer, it seems to be a useful approach to built specific subject catalogues for example in the developing countries. As regards to multiplicity of cultures, languages and indexable terms, these places seems to be ideal laboratories to see how SAP-indexing can be used to enrich conventional subjects.

References:

Atherton, P. (1978). Books are for Use. Final Report of the Subject Access Project. Syracuse University, School of Information Studies.

Cavallin, Mats (1991). Chapter Indexing. Enhancing catalogue records with chapter titles, subject titles, subject indexes and bibliographies. BIBSAM-report. ISRN KB-BIBS-R-4-SE. Office for National Planning and Co-ordination. Stockholm: Royal Library.

Cousins, S.A. (1992). Enhancing Subject Access to OPACs: Controlled Vocabulary vs Natural Language. Journal of Documentation 48(3), pp.291-309.

Ingwersen, P. (1992). Information Retrieval Interactions. London: Taylor Graham.

Ingwersen, P., Wormell, I. (1988). Means to Improved Subject Access and Representation in Modern Retrieval. Libri (38), pp.94-119.

Poulsen, C. (1990). Subject Access to New Subjects, Specific Paradigms and Surveys: PARADOKS-registration. Libri (3), September, pp.179-202.

Wormell, I. (1985). Subject Access Project - SAP. Improved Subject Retrieval for Monographic Publications. Ph.D. Thesis. Lund: Lund University, 147 p.

Fig. 1 The contents and portion of the index prepared for input.

CONTENTS Selected: 15CT

INDEX Selected: 29IT

Fig. 2 Sample of SAP database record on 3 RIP

FORMAT 1

LIBRIS
Catalogue
record

{ Post: 99   Database: LUB
{ 0-444-41329-4
{ GREENBIE, BARRIE B
{ DESIGN FOR DIVERSITY PLANNING FOR
{ NATURAL MAN
{ IN THE NEO-TECHNIC ENVIRONMENT. AN
{ ETHOLOGICAL APPROACH
{ AMSTERDAM
{ 1976

FORMAT 2

Post: 99   Databas: LUB
0-444-41329-4
GREENBIE, BARRIE B
DESIGN FOR DIVERSITY PLANNING FOR
NATURAL MAN
IN THE NEO-TECHNIC ENVIRONMENT. AN
ETHOLOGICAL APPROACH

CONTENTS TERMS

Terms selected
from the contents
page by SAP-indexing
method

THINKING

{ CT
{ THE CONCEPTUAL FRAMEWORK (3-28)
{ ETHOLOGY: A SCIENCE FOR PLANNING AND
{ DESIGN (16-28)
{ ANIMAL STUDIES (27-54)
{ TERRITORY, AGGRESSION AND SOCIETY (27-34)
{ THE POPULATION PARADOX: MORE IS LESS (35-
{ 44)
{ TIME AND THE TIDES OF LIFE (45-54)
{ HUMAN STUDIES (55-78)
{ BRAIN EVOLUTION: FROM FEELING TO

{ (55-64)
{ SPACE, SYMBOLS AND HEALTH (65-78)
{ HUMAN ENVIRONMENTS (79-180)
{ THE PARADOX OF SCALE (79-97)
{ NATIONAL SPACE (98-119)
{ OPENING UP THE SUBURBS: CLOSING UP THE
{ CITY? (120-140)
{ SOCIAL EVOLUTION (141-154)
{ A NEW LOOK AT NEW TOWNS (155-180)

INDEX TERMS
Terms selected from
the back-of-book
index by SAP-indexing
method

{ IT
{ AGGRESSION (27-34)
{ BEHAVIOUR
{ BOUNDARIES STUDY, SPRINGFIELD, MASS.
{ 169-177)
{ BRAIN REPTILIAN (56-64)
{ CALHOUN (35-44)
{ CHANGE
{ CITIES
{ CITY
{ CLASS STRUCTURE
{ COMMUNICATION
{ COMMUNITY
{ CONCEPTS
{ CONCEPTUAL
{ CONFLICT
{ CROWDING AND DISEASE (67-73) OF MICE (35-
{ 40)
{ CULTURAL

Fig. 3 Example of focused SAP strings from the Swedish SAP text database on 3:RIP software

Entered search concept(s)
In English: GREEN REVOLUTIONS
In Swedish: GRÖNA REVOLUTIONEN

       91-7058-000-6
       Miljömord eller utveckling?
       Curry-Lindahl, Kai
CT: Mat för morgondagen och den "gröna revolutionen" (395-400)

       99-0168537-0
       Dina barnbarns värld
CT: Här föddes den gröna revolutionen (28-34)
IT: Indien; den gröna revolutionen

       91-5280-141-1
       Åter till verkligheten
       Hubendick, Bengt
CT: Den gröna revolutionen (146-153)

       0-534-00347-8
       Living in the environment
       Miller, G. Tyler
IT: Green revolutions (140-146)

Explanation:
    4 different documents are retrieved, each with back-of-the-book index term phrases (IT:) or/and chapter titles (CT:) containing the entered search concept in its natural context, incl. page no. e.g. (28-34). For each document the following bibliographic data are displayed: First line: ISBN; second line: Monographic title; third line: Document author.

Fig. 4 The construction of entries for graphs by SAP-indexing

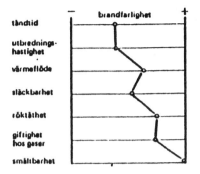

Figur 7.3 Brandfarlighetsprofil för möbeltyg, akryl.

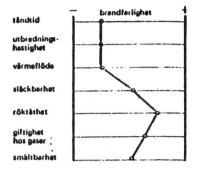

Figur 7.4 Brandfarlighetsprofil för möbeltyg, PVC-belagd bomull.

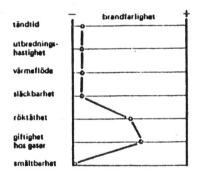

Figur 7.5 Brandfarlighetsprofil för möbeltyg, ull.

In the **SOU 1978:30** on page 122 there are 3 graphs showing the properties of furnishing fabrics subjected to fire damage *(Brandfarlighedsprofil för möbeltyg* in English means *"Properties of furnishing fabrics subjected to fire damage").* To avoid the reiteration of identical words, this set of graphs are combined in one index-string, where @ ; are used to mark relation between phrases linked together.

Thus, we have concentrated the contents of three graphs in one entry:

DT (=Diagram Text)

@ **Properties of furnishing fabrics subjected to fire damage:** **(122)**
 ; acryl ; PVC-covered cotton ; wool

Henryk Rybinski, Mieczyslaw Muraszkiewicz
ICIE, Warsaw, Poland
Gerhard Budin, Christian Galinski
INFOTERM, Vienna, Austria

The Environment Macrothesaurus System - MTM 4.0

Abstract: A new version of the multilingual thesaurus software is presented. The underlying assumption is flexibility and versatility regarding the type and the number of languages supported. The problems of merging a number of thesauri into one are discussed.

1. INTRODUCTION

Knowledge representation and knowledge engineering have been recognized by various information technology oriented communities as vital aspects of their activities. Practically, any computerized information system project includes a component related to the terminology establishment and its efficient use (Dik, S.C. [1987]). A thesaurus is especially a robust linguistic tool for a certain class of information systems and/or databases (e.g., Ciampi, C. *et al.* [1985], Fameli, E., Nannucci, R., Di Giorgi, R. [1983], Weihs, E. [1981]). This is also valid in the context of international information systems and international communities. In particular, the problem becomes tremendously important when multilingual information systems are implemented. The needs for multilingual thesauri will be growing with the development of international hyper-text products like international dictionaries, terminology vocabularies, etc.

The conviction of the information science researchers and practitioners that growing power of personal computers can be helpful in solving problems caused by the information explosion is a commonplace. Simplicity in redefining and extending character sets are additional hints in indicating microcomputers as appropriate tools for information systems development in the international environment. On the other hand high quality of information systems and databases without sacrifying simplicity is one of the "hottest" topic in the production of software for microcomputers.

Experience indicates that establishment of a multilingual thesaurus is not an easy task. Collecting the items in a multilingual environment is much more difficult than for one language. Moreover, the management and maintenance of multilingual thesauri require more sophisticated tools and skills. The process of multilingual thesaurus building is usually iterative and is a result of a consensus established among the subject specialists.

One of the methodologies in building multilingual thesauri consists in creating a new thesaurus on the basis of an experience gained with exploitation of various existing thesauri. In such a case the existing thesauri are used as a platform for the new thesaurus. In addition one can observe recently a tendency of integrating existing information communities and system by means of standardizing tools that are used by the systems.

On the other side there are no many ready-to-use software tools to handle thesauri [Ritzler, 1991]. In Rybinski, *et al* [1993] a software for building multilingual thesauri has been presented. The software known as MULTHES/ISIS has been designed as a configurable system assisting a user in creating concepts, linking them by means of a set of predefined relations, and controlling the validity of the thesaurus structure. The software has shown valuable features in building multilingual thesauri. It was successfully used for building CEDEFOP thesaurus. It is also used for maintenance of the OECD Macrothesaurus (known as MTM3.0). The main restriction, however, was a lack of tools supporting the methodology of merging essential parts of existing thesauri into one.

In this paper we present tools which allow one to work with a number of thesauri, viewing them simultaneously, and creating a thesaurus as a result of merging essential material from existing thesauri. The previous version of software MULTHES/ISIS (Rybinski *et al* [1993] was taken as a starting point in developing the current one.

The new system is based on MICRO-CDS ISIS (UNESCO [1989]). The CDS MICRO ISIS package is a widely used text retrieval software. It covers essential features of DBMS packages for PCs and minis, it is basically tailored to text databases of variable length and heterogeneous structures. Its main advantage is that it enables one to design applications without prior programming knowledge.

The main features of the MTM 4.0 software are, *inter alia,* the following:

- thesaurus maintenance and support system;
- KWOC and full tree representation and navigation tools available on-line;
- KWIC, KWOC and full tree printouts;
- defining and customization of up to 100 conceptual relationship types;
- management of facets, codes (top classification), sources, regional variants, historical notes, etc.;
- support of the various types of authority files;
- computer assisted merging; thesauri comparison by means of windows
- support of the various alphabets;
- support of linguistic and orthographic variants;
- sorting facilities consistent with national standards;
- variable length data handling;
- flexibility in defining input and output forms.

From the terminal user standpoint MTM4 fulfills the following criteria:

- user-friendliness when entering, updating, deleting, merging, checking data;
- intelligent prompting of the end user whenever in doubt;
- powerful validation facilities covering proper structuring of a thesaurus (e.g. maintenance of relationship isomorphism between languages);
- features for documenting ("keeping track") the history of the thesaurus evolution;
- availability of data protection facilities;
- availability of self-training and demonstration facilities;
- provision of a thesaurus publishing facilities at the professional level;
- modularity and openness to the further development.

In addition, MTM 4.0 should contains an Office Automation Module to cover a large portfolio of in-coming and out-coming documents and events which usually accompany the thesaurus building and/or operation.

The paper is composed as follows: the next section presents some basic notions concerned with the system MULTHES, and its use to build multilingual thesauri. Then, the main features of the system are discussed in Section 3. Ways of implementing particular features are also sketched in this section. Remarks on user interface are contained in section 4. The paper ends with remarks on possible further development.

2. BASIC NOTIONS

Besides standard thesaurus related terminology we are using in the paper some notions that are specific for multilingual thesauri applications. The basic information entities in the system are concepts, descriptors, and ascriptors (Lex [1987], Jackendoff [1989]). Any concept in a multilingual thesaurus consists of a number of descriptors, each being a concept representation in a given language. The concepts may be linked to each other by means of semantic relationships (e.g. **part_of, is_a, broader, narrower** etc.). Thus, these relationships are multilingual and are reflected in all languages. On the other hand, synonyms of descriptors may be defined specifically in a given language as ascriptors, so monolingual links between ascriptors and descriptors may be set up to model synonymy relationships, specific for a given language.

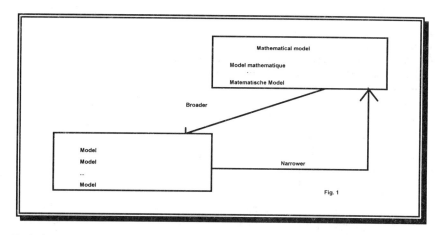

Basic data types of MTM 4.0 are terms, facets and top classification elements.

2.1. Terms
The terms take the form of descriptors or non-descriptors (ascriptors). The terms may be one of the following categories:

- generic; the terms which are specific and characterize the subject matter covered by the thesaurus;
- legal; the terms which have an official legal status and are defined in the various legal documents;
- standard; the terms which have an official status in standard documents;

- authority files; these are descriptors with special flags indicating their specific role in a thesaurus. It is recommended to store them in a special, substantive table. They should be also visible in the main data table. The authority files supported by MTM 4.0 are as follows
  - geographic; here there exists a further subdivision: political regions, geographic regions, countries, towns, lands, districts, rivers, mountains ranges, mountains, oceans, sees, lakes, forests;
  - institutions;
  - languages

In addition to the above categorization a notion of Top Term will be used. A top term is a descriptor that has no BT relationship. A special function providing a table containing only the top terms will be implemented. In case of the Environment Thesaurus the estimated number of top terms is accounted for 700 - 1000.

2.2. Facets
The facets are auxiliary multilingual objects which are used for postcoordination when indexing documents and/or searching. There are no links between facets themselves and between facets and terms. The set of facets is disjoint with the set of terms. It is predicted that the number of facets will be in the range of 10 - 50.

Note that facets constitute a separate set; however, they will be used for classifying/indexing the descriptors in the thesaurus. To this end, a special field has to be added on a descriptor worksheet (for leading languages only). The facets are reachable through a special table; they are not visible in the main descriptor/ascriptor table.

2.3. Top Classification
The concept of Top Classification is proposed for MTM 4.0 in order to classify the top terms and to assist merging. The top classification records, whose number is accounted for 50 - 100, will be implemented as an authority file. The record consists of a code and the terms (one per language)

2.4. Term structure
The descriptor record consists of the following data elements:
- the concept number (record ID)
- the term (for each language)
- top classification for top terms only
- scope note, repeatable (for each language)
- historical note, repeatable
- internal note used by a thesaurus builder only
- source
- remarks

A similar structure is foreseen for ascriptors, except for classification and scope note. In addition ascriptors may be monolingual or multilingual.

2.5 Relations
Relations within a thesaurus link the terms. Graphically, the terms and relations constitute a graph. It is worthwhile to mention that MTM 4.0 allows polyhierarchy which means that a "son"

object can have more than one "father" object. The establishment of relationships between terms and assigning them a certain semantics (meaning) is a part of the thesaurus creator responsibility.

MTM 4.0 provides a set of classic pre-defined relations such as *broader, narrower, etc.* and gives to the user's disposal a facility to define his/her own relations which might reach the number of up to 100. The nature of the latter entirely depends on the user. It is up to the user to define and name the needed relations. The MTM 4.0 pre-defined relations are as follows.

Internal
BT, NT, RT, US, UF, LV (for local variants), OV (for orthographic variants), US+, UF+

External
CR (concordance relation) which links equivalent terms from different thesauri.

One of the main difficulty of thesauri building and maintenance is a number of links that have to be established. A usual method to simplify the task is that only one-way links are to be set. The opposite direction is to be set up automatically. So, if broader type relationship is established from a concept a to b, the narrower type link is automatically set up from b to a (Fig. 1).

Usually a multilingual thesaurus may be initialized from a monolingual one. Then, other languages descriptors may be added along with ascriptors. This usually may be done by separate groups of experts in particular languages. To ensure a communication between cooperating experts it is a must to have a common language for all the software installations among these groups. We call this language a leading language. It is quite obvious that selection of a leading language does not guarantee the compatibility of thesauri maintained in various institutions. However it may be very helpful when translating the leading language thesaurus into other languages.

In case of starting a thesaurus from an existing multilingual thesaurus, it is convenient and reasonable to somehow enter simultaneously all the languages available. During this process it is vital not to neglect one of the languages at hand. The languages that are about to be mandatory are called base languages. It is clear that the leading language is one of the base languages.

It might be sometimes difficult to establish all the target languages as base languages. A reason for this could be a considerable difference between the languages alphabets. Another reason might be the fact that a language does not play basic role when constructing the thesaurus. It is therefore useful that some of the target languages are considered as non-obligatory in the first phase of the thesaurus set up. They are called secondary languages.

The consistency of a thesaurus is a condition *sine qua non* of any valuable thesaurus use. Formally a thesaurus is a digraph with the concepts as nodes and relations as edges. The main task is to avoid loops in the thesaurus graph. The validation process checking the thesaurus graph is responsible for the so called global consistency. In case of microcomputers it is rather difficult to check global consistency in dialog mode. A simpler solution is to run a batch process. Another possibility is to control the so called local consistency, which refers only to a neighborhood of the modified node. The local validation has been adopted in MTM4.0. The global validation procedure is subject of research. Additional tools to control the consistency are provided in form of displays of the thesaurus in a tree form.

3. SOFTWARE FEATURES

To some extent the CDS MICRO ISIS package offers facilities that could be used to cover the above premises. Actually, the first version of a 7-languages thesaurus has been implemented at TERMNET in 1990 on the basis of MICRO ISIS. The system has been equipped with the KWIC facilities. That software was conceived as a pilot program and by definition could not deal with all the requirements, in particular it was not versatile enough.

According to the TERMNET strategy one of the most important features of the thesaurus software should be relative ease in generating various configurations of the system, adjusted to various sets of languages. This covers adjustment to alphabets and sorting rules.

Experience acquired with MULTHES gave rise to a set of premises that should characterize target multilingual thesaurus software, which could be actually exploited by information staff working in a multilingual environment. Below we discuss the main features that are implemented.

Configuring the software
Since various cooperating information institutions may use different base languages, keeping the same leading language, a need to have configurable software occurs. Various alphabets are also concerned with the configured languages. In the software being implemented, any combination of languages configured as base and/or secondary languages along with a leading language is possible.

The total number of languages supported by the software will not exceed 9. An obvious restriction on the set of languages results directly from the limited number of alphabet characters under DOS. The limitation results also from the assumption that a reasonable response time of the update operations should be achieved with the standard PC equipment.

Another configuring factor is that various thesauri have sometimes specific types of relationships to be maintained. The record structure of the thesaurus allows one to maintain up to 99 various relations. The user-defined relations preserve a built-in *relation type* which may be BT, NT, RT US, UF, though the naming may differ from an application to another. So one can define Generic Broader relationship (and consequently Generic Narrower one), Part of (Consist of) etc.

Consistency checking
The considered software provides extended facilities to check local consistency. For example, it is not possible to repeat a term in the thesaurus. In addition, the validation procedure disables linking the descriptor to itself. Another control refers to checking types of terms to be connected. So, it is impossible to link two ascriptors by UF or USE relation. It is also impossible to connect two descriptors by the USE or UF relation.

Keeping track of the history
Two kinds of facilities are designed to support keeping track of the thesaurus evolution, namely: usage of the LOG file and provision of a scratch pad field attached to every term record. When updating, the user is prompted by the system if he wants to log changes in the LOG file. The remarks dealing with a given term may be added incrementally to the same LOG file record, or a new log record may be started. The same possibility is available when the LOG database is accessed from a standard ISIS menu. To avoid erroneous term specification, it is possible to

148

display a table for a given language and pick up the requested term, or exit and enter a new log record.

In addition to the manual LOG the system maintains automatically a log file that keeps track of every update transaction in the thesaurus. This makes possible to implement the UNDO operation with an arbitrary retrospection range. Moreover, the recovery procedures are implemented, which use the automatic log to reconstruct the older version of the thesaurus and control changes step-by-step.

Another possibility for keeping track of the changes in the thesaurus is a notepad attached to the thesaurus records. The notepad structure is definable by the user and can contain searcheable data elements.

Menu driven report generation
A reasonable number of reports may be generated by the system. For each language the thesaurus may be generated in alphabetic or systematic order. Additionally it may vary in format depending on the intended use of the printout. Special functions are designed to produce a printout in a KWIC form.

Thesauri merging
Among the new functions the merge function is of great importance. It allows one to copy some substructures of a number of thesauri into one resulting thesaurus. It is possible to observe simultaneously two thesauri at a given time. The thesauri can be viewed in various ways, e.g. by top terms, alphabetically, or in a KWOC form. So the source and target structures can be studied in depth. An example of viewing two thesauri is given in Fig. 2.

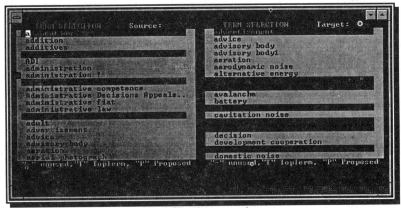

Fig. 2 Merging operation

Other features of the software
One of the main assets of the software is its adjustment to the work of distributed team of experts. It means that special features are implemented that simplify management of coming proposals. The proposals are included to the thesaurus under a special status, and before meeting of a committee they are preprocessed in a form useful for discussions. In addition special tools may be prepared for running bulk mailing concerned with the thesaurus building.

It is also worth to note that the recovery tools have been implemented, which is especially important when the large thesauri are maintained, and it is often recommended to return back to a previous state of the thesaurus.

4. FINAL REMARKS

The presented software is an advanced tool conceived to create a working background for a flexible thesaurus package. Presently the user can already adjust print formats, worksheets, alphabet's collating sequence and, obviously, one can configure the system for up to 9 languages. Especially, by defining the worksheets and record structure one can design term records tailored to thesaurus needs. Special fields may be added to keep the information on conference voting results, names of proponents, dates of proposals, etc.

A list of further extensions and improvements of the software can be set up. One of the main tasks that should be defined is to provide the software with a mechanism to detect the loops within the thesaurus graph. For the time being, it is possible to verify the structure using the function of the tree display. Another option that is planned for implementation is to link the thesaurus with the validation supported Input and Update functions.

The MTM 4.0 software has been implemented as a tool for construction of the Environment Thesaurus, being a result of merging two existing thesauri.
Multilingual thesaurus building is a multivalent and intercommunity task. Therefore, it seems that the presented software can be helpful in supporting this process. It may significantly improve the quality of traditional information systems, as well as more sophisticated multilingual knowledge based systems.

References

Ciampi, C. et al. [1985], THES/BUILD: an expert system for constructing a computer based thesaurus for legal informatics and computer low, [in:] Computing power of legal reasoning,ed. Charles Walter, New York, West Publ. Co., 1985, pp. 375-412.
Dik, S.,C. [1987]. Linguistically motivated knowledge representation, [in:] Language and Artificial Intelligence, Amsterdam, North Holland, pp. 154-170
Fameli, E., Nannucci, R., Di Giorgi, R. [1983]. Documentation in Legal Informatics and the International Bibliography on Computers and Low, Informatica e diritto, No. 3, pp. 183-239.
Jackendoff R. [1989]. What is a concept that a person may grasp it?, Mind and Language, 1989, vol. 4, No. 1/2 pp. 68-102.
Lex, W. [1987]. Representation of concepts for their computerization,Int. Classif., 1987, vol. 14, No. 3, pp. 127-132.
Ritzler, C. [1991]. Comparative Study of PC supported Thesaurus Software, Proceedings of International Conference on Knowledge Organizatıon & Terminology, Bratislava.
H. Rybinski, M. Muraszkiewicz, W. Struk, [1993]. MULTHES-ISIS: A Flexible Software for Multilingual Thesaurus Building, TKE '93, Koeln
UNESCO [1989]. Mini-Micro CDS/ISIS - Users Manual, PGI, UNESCO, Paris
Weihs, E. [1981]. Environmental Thesauri Construction, categories and function in Bavarian Land Information System, Proceedings of International Conference on Knowledge Organization & Terminology, Bratislava, pp. 104-115.

150

Ewa Chmielewska-Gorczyca, Waclaw Struk
Polish Parliament Library
Institute for Computer and Information Engineering

Translating Multilingual Thesauri

Abstract: Problems of constructing and translating multilingual thesauri based on experience gained when translating into Polish EUROVOC and the OECD Macrothesaurus (both of them covering the field of Environment). The essential difficulties in finding equivalents. Methods of overcoming these difficulties (comparison of different solutions, their advantages and disadvantages). The benefits of using the MTM program when translating the thesaurus. Proposals for improving MTM. A short state-of-the-art of the work carried on in the Chancellery of the Polish Parliament in the field of thesaurus construction.

1. Introduction

The project of translating the EUROVOC thesaurus started more than a year ago in the library of the Polish Parliament (the Sejm Library) after a survey of various indexing tools that could be used for the data bases in the Chancellery of the Sejm (Polish Parliament). Till then several different indexing languages had been in use in the various departments of the Chancellory, e.g. in the Computer Center a sort of a thesaurus (only an alphabetical list with a limited number of relations) was constructed for the data base of Polish legislative documents, in the Sejm Library's catalogue a modification of UDC, and later with the automation of the Library - Subject Headings of the Polish National Library were introduced and used simultaneously, in the press information division - in-house system, etc. After a period of research EUROVOC was chosen as the most suitable (or should I say: the least unsuitable) for general use in the Chancellery (as it covers the same field, is best suited for the needs of the parliamentary and legislative services, is familiar to the users of EC data bases exploited in the Chancellory, etc.). Unfortunately, EUROVOC was not available in computer form, only a printed version was possible. It means that besides translating it into Polish the work of typing in the English, French and German terms (as well as relations among them) had to be carried out.

The situation with translating the Macrothesaurus OECD (a few months later) was much better because it was delivered on a diskette, together with the software for maintaining and updating it (MTM - Multilingual Thesaurus Management), with the English, French and Spanish versions being available in computer form, and the German version - in hard copy (printed list). Our task was to translate it into Polish and type it with German and Polish equivalents, non-descriptors and scope notes.

As the German version was printed in alphabetical order of the English descriptors it seemed easier to translate the thesaurus term by term. In fact it appeared that much of the work had to be carried out twice because to maintain the consistency when choosing the Polish equivalents was unfeasible (uncontrollable). It means that from several Polish equivalents of one English word (e.g. for education - oswiata, szkolnictwo, edukacja) for one compound term "oswiata" was chosen, another time - "szkolnictwo", still another - "edukacja" (regardless whether in noun or adjective form). Only after typing them into the computer and grouping the into subject fields (facets or narrower terms) did the lack of consistency become apparent. Thus a time-consuming work of changing the equivalents into consistent ones had to be carried out later on.

The translation of EUROVOC was based on the subject-oriented list, and thus it was much easier to see the dictionary context of the terms (all the terms from one field), find the most appropriate meaning, maintain consistency in choosing the equivalents and avoid assigning one Polish term to two or three English ones. For example, for two English terms "hunger" and "famine" there exists only one Polish word "glod". In the alphabetical list both terms appeared in two different places and it was easy to forget when translating the second term that the Polish equivalent was already used up. When typing it the computer signalled the repetition and it was necessary to come back to the first term and find the solution.

A comparison of these two methods of translation shows that working with the subject thesaurus is more effective, it saves time, useless work and a great amount of corrections.

2. Difficulties in finding the equivalents

It can be said that in general the work of translating both thesauri was not a very difficult task. Both vocabularies use international and well-established terminology and for most of the words it was quite easy to find the Polish equivalents. However, there was a small percentage of dubious terms, and even if their number was limited, to translate them properly took twice as much time as did all the rest of the vocabulary.

The first problem was the lack of an equivalent in the target language. The reasons for that were various. For example the English term "toy library" (French: ludotheque, German: Ludothek) denotes a type of an institution which does not exist in Poland, and what is more important, we could not trace any publication on this subject in our collection, so no wonder there is no word to describe it. The same problem was with different sorts of taxes or private undertakings. Of course, a new word can be coined, e.g."biblioteka zabawek" or "wypozyczalnia zabawek" or "ludoteka" but it will convey no meaning to Polish users. It will be queer for them to see such a term, and what is more important, one can never be sure what term will be prescribed for the institution once it appears in Poland or authors start writing about it. The rules established for translating EUROVOC suggest keeping the term in the foreign version (source thesaurus version), but the inclusion of a foreign word without a prescribed meaning in a public-access thesaurus might seem bizarre to the users. It was decided that such descriptors were not to be put into the online public-access thesaurus, but should be placed on a "waiting list" (stand-by list). The moment the phenomenon (institution, object) appears, or rather the literature on it, the "waiting candidate" will be transferred onto the list of descriptors and made available to the users.

Another problem was the existence of one Polish equivalent for two English terms, as in the case of "hunger" and "famine" cited earlier. The substitute for one of those terms (or for both) had to be found showing the difference in meaning (usually by adding an adjective), e.g. for "hunger" the Polish equivalent "glod" was kept, while for "famine" - a longer term was coined - kleska glodu (disaster of famine), of course both linked by an RT relationship. Sometimes, however, it was decided to keep only one Polish equivalent for two English terms, e.g. for "computer science" and "informatics" (both of them leading to Polish "informatyka"), or "refugee" and "political refugee" (leading to Polish "uciekinier"). The result is that in the Polish version, when switching to English, users get two English equivalents. Having the Polish end-users in mind it seemed much safer not to put two separate terms with the same meaning (it might cause the doubts and critical comments).

Still another difficulty was the lack of a precise equivalent in the target language and the existence of a term with a slight deviation (overlapping) in meaning. The solution was to employ a term and to add a scope note to explain the difference in meaning (or difference in use).

The "waiting list" solution was chosen for the terms designating phenomena and objects still non-existent in Polish consciousness, reality or literature. But for some terms there was no equivalent even if the phenomenon does exist, like for "noise pollution" (and I can assure you that noise pollution is a serious problem in Poland). The reason for the absence of this equivalent is that all other terms with the word "pollution" are translated into Polish as "zanieczyszczenie" (morphologically connected with physical dirt, uncleanness, muck) and it is unimaginable to consider noise as being dirty, hence the lack of a Polish term for this concept. The simplest solution was to find the closest term representing this meaning, which was not too difficult (here: "szkodliwosc halasu" -

harmfulness of noise), but the result lacks elegancy (consistency) as can be seen from the display of English, French and Polish facets for this class.

English	French
pollution	
atmospheric pollution	pollution
chemical pollution	pollution atmospherique
industrial pollution	pollution chimique
metal pollution	pollution industrielle
noise pollution	pollution par les metaux
oil pollution	pollution acoustique
organic pollution	pollution par les hydrocarbures
radioactive pollution	pollution organique
soil pollution	pollution radioactive
(...)	pollution du sol
	(...)

Polish

zanieczyszczenie
zanieczyszczenie atmosferyczne
zanieczyszczenie chemiczne
zanieczyszczenie przemys owe
zanieczyszczenie metalami
(!)   szkodliwof ha asu
       zanieczyszczenie rop
       zanieczyszczenie organiczne
       zanieczyszczenie radioaktywne
       zanieczyszczenie gleby
                    (...)

Apart from the lack of consistency in the Polish version one can expect users to protest that "szkodliwosc halasu" (harmfulness of noise) is not a narrower term to "zanieczyszczenie" (pollution), as it has nothing to do with physical dirt. Other examples of difficulties of that kind are: conscientious objection (French: objection of conscience, German: Kriegsdienstverweigerung), telecommuting, job sharing, illegal restraint, breach of domicile, exclusion from public sector employment, programmes industry, audio-visual production, home computing, etc. The equivalents were coined (like: "ruch objecterski" for "conscientious objection), or English words kept (like job sharing) but we are well aware that they might seem a little bit clumsy to the users and are expecting the protests from their side (some we have had already).

3. Advantages of the MTM computer program

Some of the benefits of using MTM software have been mentioned already, like automatic control in assigning the equivalents, which enables to avoid two equivalents put to one source thesaurus term (elimination of duplicates). The same applies to the elimination of non-descriptors assigned to more than one term or creating non-descriptors when the same word form has been used previously for a descriptor. Both in the Polish and in the German printed version of the Macrothesaurus those cases occurred frequently, but could be perceived and eliminated only when being typed into the computer.

The comparison of both translations shows the convenience of having at least one language version of a multilingual thesaurus in computer form. Typing the equivalents, their scope notes and non-descriptors is a laborious task, but still more time-consuming and error-prone is the work of linking the typed-in descriptors by different relationships. In fact a term has to be typed as many times as different links with other terms of the thesaurus are indicated in it, regardless what kind of a relationship (BT, NT or RT) they represent. Having linked the terms in one version of a multilingual thesaurus, the work is applicable to others (it is copied automatically by computer). However, there is one advantage to linking the terms separately for every language version: typing the descriptors for the second time serves as an extra correction-tool (helps to eliminate misspelled words).

Even if praising the MTM software as a useful multilingual thesaurus tool, some minor shortcomings were revealed during its extensive use and should be eliminated if it is to be considered an effective thesaurus program. Some of the weaknesses have been signalled already to the authors and are being worked upon to have them corrected. For example, the capability to switch not only to the next screen of an inverted file, but to the previous one as well. This will eliminate typing in a term the second time if it has not been displayed as a result of a small difference in the ending (e.g. singular and plural form), and placed one line over the chosen form (unfortunately, the line not directly available to a user). There are still some problems with sorting in different alphabets, differentiation of two terms that do not vary before the twenty seventh character (which is the case with some international organizations names), etc.

Other expectations arise from the effort to make a thesaurus design more user-friendly than the traditional one, e.g. the change of a thesaurus structure and its display on a computer screen. The number of UF-terms in existing multilingual thesauri (at least both mentioned in the paper) is quite modest. Mainly the terms considered as too narrow are introduced as non-descriptors, the number of synonyms and quasi-synonyms proved to be unsatisfactory for a public access thesaurus. In the Sejm Library it was decided to put as many entries as possible, which will enable users to get to a needed concept through all existing terms representing it. As a result some of the descriptors have got ten or more UF-terms (especially in the international organizations field). In effect, when displaying the thesaurus context of a descriptor, what users get on the first screen is the list of UF-terms, which is not very helpful at this stage of a search. The most important (vital) ones for users are the BT, NT and RT terms available at the second or third screen. A solution to this nuisance is to create another display for a dictionary-context of a descriptor (as a main or additional one), in which UF-terms are removed to the end of the list (or available only when requested), and BT, NT, and RT terms cited in the direct proximity of a descriptor. Three designs of such a display are suggested in fig.1.

```
1)   BT   European integration        2)   European movement
          international security            BT   trends of opinion
     European security                      RT   anti-European movement
     NT   CSCE                                   European integration
          European arms policy                   European Party
          European defence policy                European security
          Partnership for Peace        UF   European federalism
     RT   anti-European movement             European idea
          armaments control                  European Interests
          arms limitation                    Movement
          European cooperation               pan-European movement
          European movement
          European organizations
          military cooperation
          military organizations
          strategic defence
     [ UF terms at request]


3)   OECD                         UF   Europejska Organizacja Ws...
     SN   Organization founded(...)     O.E.C.D.
     BT   European organization         O.E.E.C.
          inter-governmental            OCDE
                    organization        OECE
     NT   OECD DAC                       OEEC
          OECD IEA                       Organisation de cooperattion ...
          OECD NEA                       Organisation europeenne de ...
     RT   OECD countries                 Organizacja Wspolpracy Gos..
                                         Organization for Economic...
                                              ( ... )
```

Fig.1 Proposals for the display of descriptor articles.

The experiments in restructuring the thesaurus display require, however, further research and the testing of the reaction of users.

4. The System of Thesauri in the Sejm Library

As stated before, EUROVOC and Macrothesaurus OECD were used as the main source of terminology for the Polish thesaurus, but not without changes in the structure and scope. First of all, a polyhierarchy was fully applied to the Polish thesaurus. It means that a term can be put into more than one place in the classification scheme (subject-oriented list). Thus "Tunisia" appears as a narrower term under "Arab countries", "French- speaking Africa", "Maghreb", "Mediterranean countries", "North Africa", etc. In consequence the descriptor "Tunisia" has got in the alphabetical list several broader terms, eg.

Tunisia
BT   Arab countries
     French-speaking Africa
     Maghreb
     Mediterranean countries
     North Africa
     (...)

In EUROVOC, polyhierarchy was used only in two fields: Geography and International Organizations, and even there on a very limited scale. For the Polish thesaurus polyhierarchy was extended to all fields and used quite generously.

Another innovation was the application of a faceted structure whenever possible, e.g.

for the countries
          (by political system)
               capitalist countries
               socialist countries
          (by level of development)
               developed countries
               developing countries
               less-developed countries
          (by economy characteristics)
               agricultural countries industrialized countries
          (by religion)
               Christian countries
               Islam countries
               Protestant countries
          (by language)
               English-speaking countries
               French-speaking countries
               Spanish-speaking countries
          (by culture)
               Arab countries
     (...)

In EUROVOC all narrower terms were arranged alphabetically in "word blocks" under a heading, so that terms distinguished according to one characteristic of division were distributed, e.g.

economic development
(!)   developing countries
       development potential
       economic disparity
       economic growth
(!)   industrialized country
       integrated development
(!)   less-developed country
(!)   new industrialized country
          (...)

Testing EUROVOC while indexing the Sejm Library's collection showed that the specificity of vocabulary in certain domains was too low (especially for the legislative documents). It was broad and specific enough for international affairs (politics, cooperation, economy) but a certain number of descriptors had to be added for internal affairs (connected with political life of a country, particularly a postcommunist country in a transitional period). With the growth of the vocabulary the handling of it became a problem, the search for a term (regardless whether in computer or printed form) was more time-consuming and the control of the whole scope - more laborious. Still more serious problems were caused by different relationships for the same concept in different fields. The consequence of the adoption of polyhierarchy is that one term can appear in two (or more) places, with the effect that the same term can have different broader, narrower and related terms in each field. Sometimes the relationship between two terms is reverse and thus not acceptable by any thesaurus construction software, e.g. in the law field the descriptor "family law" is considered as broader than "family" but in social policy it is considered as narrower:

law	social policy
family law	family
family	family law

Taking all those problems into consideration it was decided to divide the thesaurus into ten separate microthesauri, compatible among one another methodologically and structurally. It means that a term can appear in more than one microthesaurus, with its form, scope note and prescribed set of descriptors being identical; the only difference can come about in the set of broader, narrower and related terms. To master efficiently ten thesauri one has to be equipped with a supplementary merging-software (or rather navigating software), so one can only hope that the further development of the MTM will provide us with the necessary tool.

5. Advantages of Multilingual Thesauri

All microthesauri are multilingual (with the English, French and German equivalents), but the foreign language versions serve as a mapping tool rather than separate thesauri. They show the terms representing a given descriptor in EUROVOC or Macrothesaurus OECD. It will enable a user, when finding no document (or not a sufficient number of them) in the Sejm Chancellery data bases to switch into the data bases using EUROVOC or Macrothesaurus OECD available online or on CD-ROM-s, and to continue the search there. Additionally, the multilingual option can serve foreign users as a main source of descriptors, which will become more important with online public access expected to be implemented in the near future.

Another benefit of having a mapping tool of that sort is the possibility to exploit external data bases when indexing foreign literature. The automatic copying and translating into Polish of the EUROVOC and Macrothesaurus descriptors used in those data bases to represent the content of documents will save time and the human intellectual effort needed for this task, which will then have already been accomplished by others.

Utterly unexpectedly the multilinguality of thesaurus proved to have one more advantage; it is quite helpful when indexing foreign literature (in the Sejm Library 50% of the whole collection), especially for those indexers that are not fluent in all of the languages covered. Thus, besides being a mapping tool it serves as a multilingual dictionary.

The further development of the system of thesauri will rely simultaneously on the growth of the vocabulary, the enrichment of existing descriptors with additional information and the evolution of the MTM software.

Roland W. Scholz and Peter Frischknecht
Department of Natural Environmental Sciences, ETH Zürich, Switzerland

The natural and social science interface in environmental problem solving

Abstract: The objectives and the curriculum of natural environmental scientists education at the Swiss Federal Institute of Technology are introduced. One major objective is the so-called *environmental problem solving ability*. The environmental scientist should be able to analyse the water, the soil and the air system, and to understand these systems and their interactions and the relationship between human beeings and their biotic or abiotic environment. For this an integrated and genuine interdisciplinary analysis of the natural and social systems is indispensible. The case study methodology is introduced as a method of teaching and research within this framework. Referring to the 1994 case study "Perspectives of the Grosses Moos", the principles of fusing methods and concepts and of knowledge integration from different disciplines are described.

Many environmental phenomena or problems evidently require reasonable treatments and decisions. The traditional disciplines and sciences, however, do not provide sufficient and adequate knowledge for environmental problem solving and decision making. This is why the Department of Natural Environmental Sciences was founded in 1987 at the ETH Zürich. One of the main objectives of research and education programs is to develop a *"General Environmental Problem Solving Ability"*. Starting with an analysis of the water, the soil and the air system, the environmental scientist should be able to investigate and to understand these systems and their interaction as well as the relationship between human beings and their biotic or abiotic environment.

From an environmental scientists' point of view the present changes and stability factors of the soil, water and air system can only be completely understood on the one hand in the frame of its social determinants. On the other hand, individual and social development are often increasingly determined, shaped and limited by natural constraints. For environmental problem solving, the interface of natural and social sciences thus is crucial. This interface is characterized by an integration and fusion of methods and concepts from the two branches of science (1).

We will report on the organization of the curriculum developed by the Department of Natural Environmental Science and on the case study approach, as currently organized by the *chair for environmental sciences: natural and social science interface.*

THE CURRICULUM IN ENVIRONMENTAL SCIENCES

The first two years: The program starts with a broad basic education in mathematics, physics, chemistry, biology, computer sciences and earth sciences (Figure 1a). So far, the curriculum is very much shaped like many courses in medicine or other natural sciences. It should be noted, however, that most lecturers endeavour to reorganize the traditional natural science curricula as far as possible in order to meet the requirements of a general environmental system view. Thus, the new Department Guide (2) denotes this part of the educational program as *"System-oriented Basic Education in Mathematics and Natural Sciences"*. The teaching of the calculus, for instance, does not begin with defining sets, real numbers, functions etc., but rather relies from the very beginning on a close to reality modelling of system change. The system view is also incorporated into an integrated basic practical training course which takes about ten hours per week through the first two years. This practical training covers all natural science disciplines and includes critical reflexion on the quality of data and methods.

Furthermore there is a series of six courses in environmental systems ranging from emphasis on matter fluxes and dynamic systems to environmental history.

With respect to education in *"Social Sciences"*, the student starts his/her systematic training with two courses in economy and law. It must be noted that, according to the tradition of engineering-schools, the student has to attend about thirty lectures a week.

Although faced with the problem of time contraints, we have recently added a major new type of training course to the curriculum. A series of lectures and exercises introduces strategies and techniques to cope with "ill-defined" problems and qualitative system analysis. This is done by presenting small case studies. Simultaneously, the student is trained in communication and presentation techniques. After this course, each student is expected to participate in a block-seminar. In this seminar, a prototypical complex and real environmental problem (i.e. a case study) is presented. One case study already presented was "The upper Rhine valley: power plants vs. riparian-forest".

This new course has been designed to single out those students, who are unable or inept to acquire the environmental problem-solving ability desired. We ascertain the presence of this ability by a specific test. The student is asked various questions concerning the case study presented in the workshop. The following skills are tested in oral and written examinations: the ability to reduce complexity, to identify relevant issues, to formulate questions that can be answered by scientific analysis, to analyse dynamic and qualitative system behavior, and skills of communication and presentation .

The final three years: After the second year, the student specializes in one discipline of natural science. As illustrated in Figure 1b, the student may choose one out of five subjects. Selecting a discipline, however must be tied to an environmental system. There is just one exception, as the environmental system may be substituted by "mathematical modelling methods". Lectures and seminars are offered with respect to both dimensions, discipline and system.

The specialization does no seek to copy the traditional type of curriculum, but tries to develop an innovative education. The chemistry education provides an example: Whereas traditional chemistry is predominantly concerned with the development and synthesis of new compounds, environmental chemistry emphasizes the fate processes of chemicals in the real world matrix in connection with the analytical characterization and quantification of chemicals. The student is expected to acquire the theoretical and technical background for an evaluation and analysis of chemical processes within the environment. Training in the *system* x *discipline matrix* takes a total of about two years.

The natural science education is accompanied by a training in the *social sciences* and *environmental engineering/technology*. The social science and humanities education covers 16 lessons a week and hence about 50% of a semester. The student may specialize in one of four areas: philosophy/ethics of science; psychology and sociology; law, economics and politics; media and environmental didactics. Like the natural science educations the lectures in the social sciences are tailored to the necessities of environmental problem solving.

At the bottom right of Figure 1b, you will find four additional elements of the curriculum. The *vocational training* focusses on developing the applied aspects of the environmental problem solving ability. The student should gain insight into the constraints of commercial environ-mental activities like planning, management and consulting.

The *environmental engineering education* is organized similarly to that of the social science education and covers a total of eight hours a week through a semester. The student may choose between different fields, e.g. forestry, energy systems, safety and risk, agriculture etc. The objective of these activities is to have the students experience environmental problems in the frame of technical and socio-economic constraints.

The *case study* is a new type of course and covers 18 hours a week during the eightth semester. As it is a most comprehensive innovation of teaching and knowledge organization with respect to environmental science, we will report on it in more detail.

158

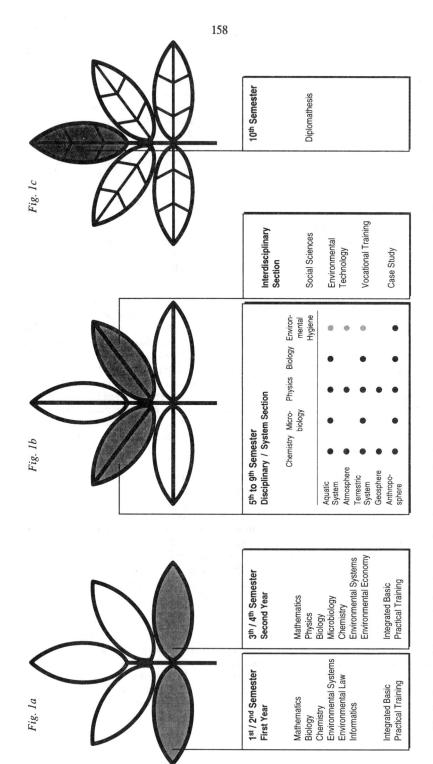

Fig. 1c

Fig. 1b

Fig. 1a

1st / 2nd Semester
First Year

Mathematics
Biology
Chemistry
Environmental Systems
Environmental Law
Informatics

Integrated Basic
Practical Training

3th / 4th Semester
Second Year

Mathematics
Physics
Biology
Microbiology
Chemistry
Environmental Systems
Environmental Economy

Integrated Basic
Practical Training

5th to 9th Semester
Disciplinary / System Section

	Chemistry	Micro-biology	Physics	Biology	Environ-mental Hygiene
Aquatic System	●	●	●	●	◐
Atmosphere	●	●	●	●	◐
Terrestric System	●	●	●	●	◐
Geosphere	●	●	●	●	
Anthropo-sphere	●	●	●	●	●

Interdisciplinary
Section

Social Sciences

Environmental
Technology

Vocational Training

Case Study

10th Semester

Diplomathesis

Fig. 1: Course of the curriculum in natural environmental sciences.

THE CASE STUDY APPROACH

The *case study* is considered both as a *method to learn,* to teach and to transfer knowledge as well as a *research methodology.* As a teaching strategy, the case study approach has been developed at the Harvard School of Business Administration. As a research method it particularily permits a scientific treatment of vague (i.e. „ill-defined") and complex real world problems and has first been applied in anthropology and the social sciences (3, 4, 5). We regard case analysis as a suitable paradigm to attain insights into the complexity of environmental decision making and also as a method of *integrative decision* making. Doing case analysis provides insight into the process of acquiring and integration and into qualitative aspects of environmental decision making. After describing some principles of the case study approach, we will demonstrate how the natural and social science interface is being organized within the current case study "Perspectives of the Grosses Moos".

Principles of case study research: Like other research methods the case study approach starts with understanding the prior change of a well defined system. Besides, a case study analysis should meet (among others) the following requirements:
a) Description and analysis of the case should be „complex, holistic, ... involving a myriad of not highly isolated variables from natural and social systems. Furthermore the data are likely to be gathered at least in part by personalistic observation..." (6, p.8).
b) Within the current social and natural sciences there is a strong tendency on favor of generalization by statistics. In contrast, the researcher applying the case study approach should rather generalize on the basis of analysis rather than data. This principle should hold for both generalization and abstraction in the case analysis and in generalization from the case (7).
c) Compared to traditional research the relationship between the researcher and the object of research is different. We do not speak of subjects and objects, but rather of participants and agents of the case study. In order to attain a holistic perspective, from an epistemic point of view, it is necessary to integrate the knowledge of the persons most closely acquainted with the system.

Types of Case Study design: There are various types of case studies. One may discriminate between single case and multiple case and between holistic and embedded (single vs. multiple units of analysis) case studies (3, S. 46).

Objectives of the study: The Grosses Moos is a former marshland within the triangle formed by three interconnected lakes west of the capital of Switzerland, Berne. Between the three lakes there is a wide plaine of about 500 square kilometers. There are 30 communities with a total of 42.000 inhabitants in that area.

The Grosses Moos was been repeatedly drained in the last 130 years in the course of huge melioration programs. Today, it is most intensively used by vegetable gardening. Several major environ-mental problems have arisen with respect to overfertilization, subsidence of the drained marsh, biodiversity, the aesthetics of the landscape etc.

The problem or principal question and the initial objective formulations in the 1994 case study "Perspectives Grosses Moos" can be summarized as follows:
- Which decisions and changes must be made with regard to the future use of agricultural soil in the Grosses Moos to attain a sustainable development of the entire regional ecosystem, within the existing legal frame?
- What can the Article 31b of the Swiss Agriculture Act contribute in particular?
According to this new article, farmers will receive a direct reimbursement for areas cultivated in accordance with the established rules of biological farming. To obtain these funds, the farmer also must re-nature 5% of his farmland area.

CASE STUDY '94

Future use of the agricultural area 'Grosses Moos', in accordance with the legal frame aimed at sustainable development. Perspectives and consequences for nature and society.

1. ECOLOGICAL ASPECTS	2. ASSESSMENT OF AGRICULTURE	3. ECONOMY AND POLITICS	4. SOCIAL ASPECTS
1.1 Regulation of the water system: Artificial changes in the hydrological system	2.1 Agricultural diversity: Interrelation between the diversity of type of cultivation	3.1 Agricultural economics: Perspectives of the 'Grosses Moos' within the agricultural commodities market	4.1 Population: Ecological attitude; acceptance of ecological laws
1.2 Canals and lakes: Close to natural realisation of the canal system	2.2 Areal aspects: Interrelation between field size and ecology	3.2 Other branches: Economy and ecology	4.2 Information and education: Efficiency of programs for information, advice and education
1.3 Other areas: Ecological potential of the settlement area and of the agricultural area	2.3 Energy balance: A tool for the ecological assessment of the agriculture	3.3 Administration: Implementation and control of agricultural / ecological laws	4.3 Change of social values: Conflicts between amelioration programs and change of social values
1.4 Wind protection strips, hedges and forest: Proposals to enhance the ecological status	2.4 Nutrient balance: Comparison of different types of cultivation	3.4 Interest groups and parlaments: Agriculture politics within the 'Grosses Moos' region	
1.5 Soil development: Conflicts about soil use	2.5 Pesticide, Herbicide etc.: Dynamics of regional matter fluxes		
	2.6 Agribusiness: The effect of different types of cultivation		

Fig. 2: Design and contents organization of the case study.

The Grosses Moos may serve as an example for a region where the principle problems cannot be solved by an immediate, clearly defined technical action plan or decision. Usually, problems of the Grosses Moos type have been banned from research due to their complexity and vagueness. The choice of the topic Grosses Moos was motivated by the objective of finding a prototypical example for complex environmental problem solving and to optain some experience with the natural and social science interface.

Planning the study: In this case study, 105 students belonging to the eightth semester partici-pated. The study was prepared by a commission consisting of 15 students and two professors (Profs. Koller and Scholz). The first preparation task was to develop a design of the case study. We took an embedded single-case design (4). We have considered just one object, the Grosses Moos, while making allowance for multiple perspectives and multidisciplinary analysis. The commission also prepared agendas for the case study, for the four projects and their 18 subprojects.

Interdisciplinarity and knowledge integration in the planning: The design and the content-organization is displayed in Figure 2. It is structured into four systems, i.e. ecology, agriculture, economics and politics, and the social system. For each system, there are various subperspectives or subprojects. These subjects present the disciplinary perspectives such as hydrology (1.1), pedology (1.5), economy (3.2), social psychology (4.1) or projects in which several disciplines have to be applied simultaneously. Subproject 2.6 and 3.1 require both agricultural science and economics. During the case study, four or five students cooperated in each subproject.

It must be noted that the case study point of view (see above) not only requires integration of disciplines and systems. The knowledge of the different participants has to be synthesized as well. For this purpose, we established counceling boards with farmers municipalities, public authorities and interest groups from the very beginning.

This cooperation was established for two reasons:
- First, when integrating the different knowledge that stakeholders and the people who live and work in the area may offer, the "goal of competence" of the case study agents can be met (8). This competence will be used in the process of finding a problem solving strategy based on a common consent.
- Second, when looking for a solution to an environmental problem, the people in the area are crucial, as in the end it is their share to realize problem solving strategies towards a sustainable development.

Thus, the process of *knowledge acquisition* in some respect has been organized as a three dimensional bootstrapping with the dimensions:
- *disciplinary organization*, i.e. bridging the natural and social science interface,
- *system view* and
- *type of knowledge present in the different agents* of the case study.

Knowledge acquisition: As the researchers (i.e. the students) were not sufficiently prepared with respect to all their tasks in the subprojects, they used three weeks to acquire a broad knowledge. About fifty lectures, workshops and excursions were arranged with specialists from various disciplines and "system experts". These arrangements built the *first stage* of our case study and provided basic orientations with respect to the Grosses Moos, the various environmental systems and the different disciplines involved.

The *second stage* took about two months. The subjects were confronted with the problems posed to the various subprojects. In each of the eighteen projects the students had to investigate the Grosses Moos with respect to their agenda and to the general objectives of the case study.

Knowledge integration and the natural and social science interface: The students were confronted with the natural and social science interface already existing in subproject work. To give an example: one of the tasks of subproject 1.1 was to reconstruct the man-made drain and canal system in its historical course. The participants were called to identify among other things the various interests, i.e. agriculture, fishery, soil protection, flood protection, power plants, tourism and the groups representing them. Each interest group formulated different requirements for the lake levels. The hydraulic system is determined by social interests. More general it would be most interesting to find out how the hydroengineer knowledge is organized around these interests and where environmental issues are being placed.

There were some more interesting findings with respect to hydrology. There are a host of methods and models to determine how much water will flow into a canal depending on the area tied to the canal, the type of cultivation, on the gradient of the area, on the type soil etc. These models, however were not constructed for the application to the soil used by vegetable gardening. Hence, the hydrologists had to use the method of oral expert interviews to assess the quantity of water in the canals.

The main objectives of the case study were to develop an understanding of the system and a midterm basic orientation toward a sustainable development of the Grosses Moos. To cope with this question, the knowledge attained in the various subprojects had to be integrated. Synthesis and knowledge integration was the *third stage* of the case study. It was organized into different groups and covered the final three weeks of the study. The most comprehensive knowledge integration was conducted in the "Area Negotiation" and in the "Szenario Analysis" group.

The "Area Negotiation" group had to develop a problem solving strategy for an optimal conception and design for the 5% of land to be be re-natured of the entire area of the Grosses Moos. You can imagine that this approach is full of inferential dilemmas and social conflicts. The latter are particularily caused by the potential requirements to swap property in order to interlink ecological zones.

We will briefly scetch the synthesis-work of the Szenario-analysis group. The objectives of this group were extraordinary close to the initial goal of the case study as a whole.

The *first step* consisted in achieving a conception of the general goal of our case study based on common consent. This was a redefinition of the concept of sustainability with respect to its ecological, economical and societal dimensions. Note that sustainability itself is a concept, which is much more on the holistic side, although we tried again to gain insight and understanding by an analytic decomposition, discriminating between ecological, economic and social sustainability. Based on this decomposition, we even identified critical variables which will be used to evaluate different szenarios.

The *second step* dealt with determining the major variables affecting the logic, development and change of the Grosses Moos. In order to construct a myriad of not isolated variables from the natural and social system, a table of 24 impact variables was created in an iterated group-process (10). For all these variables written definitions have been developed and agreed upon in the team to establish a common terminology. Then the variables, believed to sufficiently re-present the system and its dynamics, were intuitively selected.

The *third step* of our journey into knowledge integration was to construct a cross-impact matrix. According to the method of szenario analysis (11), we tried a semi-quantitative approach. The direct cross-impacts of the various variables were rated on a three-level scale having the levels "no impact", "medium impact", "strong impact". This process rendered a deep insight into the structure and dynamics of the system.

It must be noted, that we rated the current degrees of direct impacts. Indirect impact, e.g. if variable x affects variable z via y, were not considered and rated. Also, the relative importance of the different variables was only determined by degrees of influence and not by other characteristics.

Even this rough analytic cross impact matrix allows of operations on the sign level. If we accept the fuzziness in the change of representations, we may attain further insights into our

163

conception of the system by formal operations. For instance, if we simplify this cross impact matrix to a matrix A having only two degrees of impact, say 0 for no or small and medium impact and 1 for strong impact, matrix algebra might be helpful as it conceives of the system as of a discrete automate. By multiplying matrix A with itself, according to the method of MICMAC-Analysis, we are able to attain information by the matrix A^n about the order of the 24 impact variables according to the extent they are incorporated into the network of impact variables.

It is most interesting, that this analytic mathematical modelling using an intuitively generated group consensus about the cross-impact of the relevant variables, serves us to derive evidence that some hypotheses disputed within environmental science may be valid. In particular, we were able to substantiate that environmental variables like degree of interlinkage of re-natured areas, type of cultivation, fertilizer use - within a 20 year perspective - are of minor importance. They are less embedded and interlinked within the network of impact variables compared to variables of social systems like laws and interest groups.

CONCLUSIONS

There is a special need for an adequate conception of the natural and social sciences interface with respect to environmental problem solving. The Depeartment of Natural Environmental Sciences of the ETH Zürich has incorporated a program of social science education into the natural environmental science education. Within the framework of environmental sciences, both the natural and social science have to be strongly reorganized in comparison to the curricula of the traditional disciplines. With respect to teaching and research there are two approaches which may help to develop the natural and social sciences interface. First, there is the system view and second the case study methodology. With respect to both approaches, knowledge integration is necessary and crucial and may be organized in a similar manner as proposed already by the Brunswickian lens model (11). In decomposing a problem with respect to different disciplines or aspects, models of knowledge integration are applied. Though system analysis and szenario analysis provide some tools, we certainly need more techniques, which help to organize the interface between the natural and social sciences as well as to integrate intra- and interpersonal (cf. 10) knowledge .

REFERENCES

(1) Scholz, R.W. (1993). Interdisziplinarität als Grundprinzip. ETH Bulletin,, 251, 21-24.
(2) Frischknecht, P.M. and U. Kumpf (1994). Wegleitung für den Studiengang Umweltnaturwissenschaften. Zürich: ETH Zürich.
(3) Yin, R.K. (1989). Case Study Research. Newbury Park Ca.: Sage.
(4) Hamel, J., Dufor, S. and D. Fortin (1993). Case Study Methods. Newbury Park Ca.: Sage.
(5) Yin, R.K. (1993). Applications of Case Study Research. Newbury Park Ca.: Sage.
(6) Stake, R.E. (1976). The Case Method in Social Inquiry. In The Case Study Approach to Educational Program Evaluation in Britain and „the Colonies". University of Illinois: Centre for Applied Research in Education.
(7) Blumenberg, H. (1952). Philosophischer Ursprung und Philosophische Kritik des Begriffs der wissenschaftlichen Methode. Studium Generale, 5, 133-142
(8) Renn, O. and Th. Webler (1992). Anticipating Conflicts: Public Participation in Managing the Solid Waste Crisis. GAIA,, 2, 84-94.
(9) Götze, U. (1991). Szenario-Technik in der strategischen Unternehmensplanung. Wiesbaden: Deutscher Universitätsverlag.
(10) Scholz, R.W., Mieg, H.A. and O. Werber (in prep.). Mastering the complexity of environmental problem solving with the case study approach. To appear in: Zimmmer A. and R.W. Scholz (eds.). Qualitative Aspects in Decision Making. Amsterdam: Elsevier.
(11) Brunswick, E. (1956). Perception and the Representative Design of Psychological Experiments. Los Angeles, Ca.: University of California Press.

Agnes Ajtay
Csaba Molnar, Tamas Toth
Eötvös Lorand University, Department of Cartography, Budapest

Education in Cartography and its Relation to Environmental Knowledge Organization

1. Introduction

Geographic knowledge plays a most important role in the environmental sciences. Cartography should therefore be looked at as a discipline of a basic character to this knowledge area. 'Cartography' has been defined as

> the art, science and technology of making maps, together with their study as scientific documents and works of art. In this context maps may be regarded as including all types of maps, plans, charts, and sections, three-dimensional models and globes representing the Earth or any celestial body at any scale (MLD, see below)

Maps have been defined as

> a representation, normally to scale on a flat medium, of a selection of material or abstract features on, or in relation to the surfase of the Earth or of a celestial body (MLD)

Besides its relationship to geography, cartography is also related to geodesy, which includes surveying, topography, photogrammetry and remote sensing, to regional planning, and finds application areas in computer science (computer-cartography), information science (information systems) as well as in public education, applied graphics and typography.

The main source of terms and definitions in cartography is the Multilingual Dictionary of Technical Terms in Cartography (MLD) which was compiled by Commission II of the International Cartographic Association ICA. Commission II was established at the Second General Assembly of the ICA, held in London in July 1964. The book was edited in 1974. Its subtitle: Definition, Classification and Standardisation of Technical Terms in Cartography. It contains
- Definitions in five languages: German, English, Spanish, French and Russian and equivalent terms in nine other languages: Czech, Italian, Japanese, Hungarian, Dutch, Portuguese, Polish, Swedish and Slovak; as well as
- Appendix A: Equivalent names for individual map projections;
- Appendix B: Equivalent terms for groups of map projections;
- Alphabetical indexes in all languages;
- Select bibliography;
- Sample sheet-layouts.

ICA is the biggest organization of cartographers. Its experts are engaged in cartography dealing with theoretical as well as also with practical questions.

2. Classification of Maps

A part of the teaching of cartographic theory is map classification. Before dealing with modern teaching programs in this field I should like to point out that within the Universal Decimal Classification (UDC) maps are treated under 528 (within the area of the Mathematical and Natural Sciences) and under 912 Geography, Biography and History. From the cartographic aspect, however, according to the Hungarian vice-president of the ICA, the UDC does not supply the necessary classes for map classification. In the following the different and most important points-of-view under which maps can be classed are listed.

According to the target:
- orientation
  topographic maps, tourist, town and road maps, maps for orientation running
- education
  supports mostly the geography and history by maps, with didactic representation of map symbols
- information
  general view on geographical, natural, political, economical situation of certain territories, mostly in small scale maps, like atlas sheets, text figures, maps in television and newspapers.
- prognosis
  representing certain situations; e.g. built-in areas, roads, railways, agriculture, etc. to show plans of shorter or longer periods of time or represent ideal imaginations.
- publicity
  themes for the public, e.g. booklets for the tourism.
- result representation
  based on original local data showing the result in a certain branch of science to reveal new connections and characteristics.
- navigation
  base for navigation in the air, on the sea and on the roads.

According to the scale:
- large scale: 1:10,000 and bigger;
- middle scale: 1:10,000 symbol 190 \f "Symbol" \s 12§ 1:300,000 cca.
- small scale: smaller than 1:300,000

According to contents:
- topographic map
- thematic map

For many cartographers all maps are thematic maps which means, they are related to certain subject areas and fields. For such a thematic classification a universal classification system is needed.

We can classify further

according to the edited form:
- one sheet
- map series
- atlas

According to the way of use:
- handy-map (maximum 100x120 cm)
- wall-map
- text figure
- atlas

According to the method of production:
- surveyed map (topographic base map, different scales in different countries)
- compiled map (map compilation from a topographic - surveyed - map.

According to the area represented:
- region (e.g. an industrial agglomeration)
- state
- group of states
- continent

- group of continents
- world

According to the form of representation:
- main map
- inset map (enlarged map detail at the margin)
- framed map face with sheet margin
- map on bleeding sheet (without margin)
- map: an island alike

According to cartographic projection:
(to be presented in the following section)

3. Modern Teaching Programmes in Projection-Teaching

at the Eötvös Lorand University (ELTE) (with software demonstration by Cs.MOLNAR and T. Toth)

In Hungary the ELTE teaches cartography for their students for five years. Here the students get some information on the process of map-making from the mathematical-geodetic basis to the printing. The curriculum contains also other sciences which results are used by cartography (e.g. geography, geodesy, photogrammetry, etc.) or which are using a great deal of maps and must know the requirements (e.g. geography, meteorology, geology, etc.). The teaching of *Projection* is one of the most important subjects because it deals with the mathematical basis of cartography.

Projection (mathematical cartography) usually causes a lot of problems for the students because of its mathematical implications. In the following we will deal with the computer-aided teaching of these subjects.

Cartography - just like the sciences - tries to use the opportunities of the computer. One can make maps and charts with computer-assistance, one can build up spatial databases, one can represent these data on a map, and use the GIS (Geographical Information System), etc. However, computers have not been connected to the teaching of cartography in Hungary yet. We think it is not worthless to study this topic: we observed that after the third year students use already efficiently our teaching programme (VETTI) in order to prepare for their exams in Projection.

In making a scholarly use of computers one must keep in mind the following basic principle: Do not use computers if the given problem can be solved without a computer just on the same level! It is worthless to show only pictures and slides or books in a typed form on the screen. But it can be very useful if one uses a computer inside the teaching environment. At the same time we think it is also useful to make teaching programmes that solve only a part of a problem or can help in the development of entire syllabi. The most proper programme also may contain unsolved problems, so the role of the teacher is to determine, to control, to help, and to consult. Our goals are still very modest.

In the following an overview is given of the different types of teaching programmes of cartography and examples from the topic 'Projection' are shown.

Traditional teaching programmes
The traditional teaching programmes work inside the frame of a common computer. They use text and graphics (sometimes also sound, but only sound effects). These teaching programmes can be classed according to function in the following way:

- knowledge transmitting,
- practising,
- representing,
- simulation,
- examining,
- other teaching programmes.

As mentioned before a good teaching programme can be of better service than a usual book or film. With respect to Projection this kind of service is, for example, the 'display' of the grid-system of different projections.

The science of Projections in its mathematical aspect concerns the relationship between the main surface (usually the surface of the Earth) and a sheet of paper. Because of the difference in the Gauss-curvature of the flat paper and the surface of the Earth there are torsions in projections. We distinguish three kinds of torsions: linear-, areal- and angle-torsion. On the base of the difference in the represented area, theme, or the demands of the user we have to choose different projections for the base map. (For example the climate-maps need parallel-running parallels on the map if we want to show the zonality properly.) Not only the different projections but inside one projection the different parameters as well cause different representation of a selected area on a flat paper. If a map maker wants to use the most proper projection for a given theme he or she has to know the bad and good attributes and views of different projections and thus needs theoretical and visual knowledge. Because of the infinite variation of parameters of projections this is a hard challenge for the students and the representation is impossible in a traditional schoolbook. The problem needs the computer. We tried to solve this problem with our masterwork called VETTI which is a teaching programme for projections.

It is hard to classify this programme: it transmits knowledge and represents it at the same time. It is a traditional teaching programme because the flow of information is done through the screen and the base elements are only text and (mostly) graphics.

The main functions are the following:
- representation of the grid-system of a selected projection with selected parameters with or without the contour of the continents
- one can see another projection over the first for comparing
- one can see torsion ellipses of a projection to perceive the torsions
- one can get information about the original, its history and the use made of the projection
- one can get information about the scientist working with the projections
- one can save the drawn graticule to the winchester
- one can represent the saved pictures again
- one can make a short demonstrative programme from the saved graticules

Demonstration VETTI

Modern teaching - Multimedia and hypermedia
There is a big disadvantage in the traditional teaching programmes: they use only text and graphics. The set of the base elements of the multimedia systems is the following:

- structured text
- graphics
- animation
- slide
- video
- sound
- music
- human voice

In a multimedia system the most significance elements are the video, the hi-fi sound and the use of the human speech.
In the case of multimedia systems one needs to develop thus kinds of materials which improve the efficiency of the studying, the level of teaching, make the learning and the long run remembrance easier. These systems are concerned with auditiv and visual channels at the same time, they use more organs of sense at the same time, therefore they make the learning easier for the students. Bayard-White observed that one can remember only 20% of listened, 50% of listened and seen at the same time, 90% of

spoken and done material. This means that one must make certain to include some activity in the student's learning process, which means, the teaching programmes must become interactive. Such an interactive multimedia is possible in using hypermedia. Hypertext is a kind of a hypermedium which uses basically text but may also work with graphics, music and video.

Hardware requirements:
Hypermedia systems exist on IBM PC and Macintosh computers. There is a multimedia standard which has - in the case of IBM PC:
   an AT 486 computer with 8 Mb RAM, 200 Mb winchester, SVGA colour display and
   audio and video tools
The input can be by keyboard, mouse (trackball), and microphone (for speaking).

Software:
for hypertext: Guide, ToolBook, HyperCard
for multimedia: AuthorWare Pro

Advantages:
-   own learning speed (interactivity, non-linear materials)
-   easy understanding, longer remembrance (more sense of organ)
-   winning time in the teaching (for the teacher)
-   self-test possibility
-   creativity
-   efficient representation

Disadvantages:
-   costs (an efficient system is very expensive)
-   resistance of the institution
-   resistance of the teachers (need time to make materials, no appreciation)
-   TIME! The solution: DSE (Direct Self Education) - students make materials for other students
-   exhausting (One must be careful with using the media! E.g.: Don't use music and sound effects without any reason)

Hypertext in the teaching of Projection
   A typical hypertext software (e.g. Guide) contains a text editor, a graphical editor, database (text, graphics, music, video) and reading possibility (Viewer) - a rather complex integrated developing environment. A functioning hypertext application is a branch of nodes and links. The node is for one occasion displayed information on the whole screen or only in a window. The connections between the nodes are called links. If one wants to step from one node to another one has to choose a link but the developer's task is to reach a well-constructed structure. So the power of these hypertext systems is not only in the given information but in the connection between the information packages as well.

The main advantage of hypertext systems is their flexibility. But how can we reach that the user follows the proper way without any reservation? There are three main rules for the application of hypertext:

-   use always the best way
-   organize the main arguments linearly
-   insert remarks into the text where the user will get additional information

Loosing a way is the main problem in hypermedia systems. The authoring systems contain some kind of navigation module to solve this problem (e.g. the user may look-up the way already being done).
Hypertext applications used in the teaching of projections can be made easily. Our goal is to make a hypertext application for the teaching of not only the basic elements of projections but the analytic science of projections as well. So we made an exact overview of different projections in the same structure.

The projections are fit into a hierarchy. In the overview we follow a well-determinated structure. The main description of a projection is a node. One can reach the following information from here:

- the representation of the grid-system in the chosen projection with the chosen parameters (graphics)
- the history of the projection with old maps drawn in that projection (text, graphics)
- biography of the developer and/or the first user of the projection (text, graphics)
- the mathematical equations of the projection
- the significance of the torsions of the projection, equations, displaying the torsion ellipses, iso-lines on the grid-system

Demonstration Hypertext in the Science of Projections
(to follow actually)

Jela Steinerova
Department of Library and Information Science, Faculty of Arts, Comenius University,
Bratislava, Slovakia

Education and Training in Knowledge Organization and Information Retrieval at Comenius University as a background to environmental Knowledge Organization

Abstract: The paper aims at developing a new approach to teaching courses on Information Retrieval and Information Consolidation through the concept of knowledge organization. Cognitive modelling and the interdisciplinary model of knowledge structure in information retrieval is described as integrated into traditional courses. The perspective of further information service modelling is shown for practical development of libraries and information (environmental) centres.
In order to break the traditional thinking in information retrieval research and teaching the transition from technological aspects to dynamics and complexity is emphasized and presented in the outline of the course core.

1. Introduction

The aim of this paper is to present the specific approach to education and training at the Department of Library and Information Science with regard to the subject of Knowledge Organization. The new curriculum at the Department approaches the area of knowledge organization from the viewpoints of information analysis and information retrieval. We are trying to link the subjects of Information Retrieval and Information Consolidation through the essence of information processing concentrated around knowledge organization. We would like, in our new concept, to emphasize the cognitive research results including cognitive processes (general and specific), cognitive abilities, cognitive skills and cognitive styles. Based on cognitive modelling we present the outline of the knowledge structure model in information retrieval. Practical implications are embodied in the proposal of the course core for teaching courses on Information Retrieval and Information Consolidation.

2. The Past and the Present State-of-the-Art

In the past ten years knowledge organization at the Department of Library and Information Science has not been developed (within the new concept), but it was partly comprised in such subjects as
Information Languages (including cataloguing and classification) and Information Retrieval Systems. But the concept of knowledge organization has been unknown until the 1990s. Then the new curriculum started to be elaborated and one of the important contents changes included the concept of knowledge organization (since 1991) within the compulsory subjects of Document Processing, Information Languages, Organization and Use of Information Sources, Information Analysis.

Another change of the curricula (since 1992) brought about even a more positive shift towards the subjects and courses dealing with knowledge organization. The contents of the compulsory courses on Theory of Information Retrieval and Information Consolidation have been revised with regard to the new concept of knowledge organization and knowledge representation. Apart from this, there is a large choice of optional courses for our students, which include knowledge organization principles from various special viewpoints, e.g. Expert Systems, Artificial Intelligence, Hypertext, Conceptual Analysis, Knowledge Representation. The major part of these courses is connected with the specialisation in Information Analysis covering several special levels (I-IV). (Document Analysis, Linguistic Analysis, Conceptual and Cognitive Analysis, Factual Analysis).

Knowledge Organization in Subject Areas, INDEKS Verlag, Vol.1(1994)p.170-178

Our concept of knowledge organization will be presented with regard to two special courses within the compulsory part of the study, i.e. Theory of Information Retrieval, and Information Consolidation (Special Methods of Knowledge Organization).

3. The Background of the New Approach

The new approach to the courses on Information Retrieval and Information Consolidation takes into account the paradigm shift in information science research. We would like to avoid the traditional simplification of the information retrieval processes marked by formalizing models with prevalent quantitative data and a positivistic point-of-view. We regard information retrieval as a complex process with special dynamics caused mainly by knowledge structure. Information retrieval deals with the change of knowledge and following this viewpoint it is necessary to concentrate research and teaching on the qualitative approach together with the concepts of dynamics and complexity. Some researchers consider these categories to be the most important concepts for new ways of investigating social systems within the frame of systems modelling (e.g.dynamics+complexity-dynaxity) (used by Löckenhoff 1994) and this is also true for information retrieval. As the emphasis is laid upon the representational systems capable of evolution, we would like to change the traditional picture of information retrieval towards its dynamics, it means towards the systems which will take into account their environment and their users in that they will evolve together with them. We agree that the deterministic, linear models are no more longer suitable for dynamic changes of information retrieval. That is why we are trying to find the basis for new modelling of information retrieval. This should be oriented towards the contents and the meaning of a message which is sometimes regarded as the knowledge itself as opposed to message and information (Jaenecke, 1994). ("A message contains knowledge (constitutes knowledge) if its contents consists in universally valid statements on the world").

Based on knowledge structure research we suppose that the new task of information retrieval consists in selection and synthesis of knowledge from the flood of messages. This is the intersection of knowledge organization and information retrieval as both sides of the information processing world. The social significance of the problem really means that information technologies contribute to the flood of messages and information, but it is still difficult to select appropriate knowledge for a special user in his special environment synthesized into the required context. Information retrieval should, at this stage of its development, concentrate on the relationships among the units of knowledge in order to avoid "pseudoknowledge" (Jaenecke 1994) and "mental pollution". In this respect it has to do with social and environmental consequences which can lead to difficulties in handling real knowledge in information systems.

Another important point for developing the course on the Theory of Information Retrieval is that the education of library and information workers has been changed in the past ten years towards more generalized approaches with prevalent orientation on users and the environment of information systems specialized on types of services (not types of institutions). What is important is the problem of the common information process regardless of formal organization. Knowledge and skills (in handling the process and the general information retrieval principles) can then be applied in a large variety of institutions in which graduates of our studies can be placed.

The usual topics of information retrieval should be completed by new concepts of creative chaos, self-organization, dynamics and evolution. Creative intuition composed of reasoning and feeling in mutual harmony has always been omitted from the models of information retrieval. Anyway, it is very important in building hidden order and representational structures which should be incorporated into information retrieval process. One of the ways of grasping these structures is a conceptual structure within information retrieval which is built especially of conceptual relationships in conceptual networks.

That is why the problem of the information retrieval course concerns - in several questions - problems and tasks of knowledge deficit regarding knowledge organization, as well as the problems of mechanisms dealing with these tasks.

The last viewpoint of the background is formed by the principles of qualitative research as explained in several works (Sutton 1993, Bradley 1993). The implications for information retrieval represent especially the stress laid upon the active, self-defining social world of human construction. In information retrieval this point is reflected by knowledge organization involving open-ended and interpretative role of information workers and users who are managing complexity as opposed to reductive abstraction and quantification. In this way we could humanize our course and its topics connected with information retrieval. The backbone of information retrieval viewed from the perspective of knowledge organization is built on the principles of applying the management of complexity and dynamics and implies especially contextualisation (an approach to social-scientific observation that takes into account the environment in which the observational event takes place), understanding, pluralism and expression (as the four major attributes of a qualitative research).

4. Cognitive Modelling

Based on knowledge structures, cognitive modelling in information retrieval represents one of the most productive ways of explaining the essence of information retrieval. Knowledge ordering is seen as the continuity of the learning process of the user in information retrieval. Cognitive objects are composed of human elements (users, mediators, authors), as well as of representational and organizational mechanisms. These objects can be described in terms of their contents, their evaluation and decision-making directed to certain paths which should be followed in information retrieval. Together with related information and the context (internal and external relationships) we (as humans taking part in information retrieval) form the meaningful wholes (following the holistic and synergetic aspects). Practical, computer modelling is in this way of thinking embodied in declarative query languages and object-oriented database systems which are proposed to process a large amount of complex multimedia objects reflecting our modelled cognitive objects.

Another complication, however, is caused by the environmental approach which takes into account a deep knowledge of the total environment within information retrieval and the likely conditions that lead to failure or success in that environment.

What is then the function of knowledge organization in information retrieval? We suppose that it is selection, evaluation and interpretation in the form of a specialized filter followed by a specialized synthesis composed of creative transformation of knowledge including new (specialized, and also personal) viewpoints and relations among units of knowledge. But it is a process (and processing) characterized by recursive mechanisms in knowledge handling which occur in interconnected, self-organizing cycles forming a spiral. As we can see, we should reassess some fundamental questions concerning process, time, space and their modelling under the conditions of complexity and dynamics. The change cannot be understood through linear means of modelling because the relationship between man and his information retrieval environment has been constantly changing. In this respect, environment (and even ecology) is the concept representing the influences of context in information retrieval and it is no longer the secondary but the primary factor that should be modelled in cognitive modelling of information retrieval.
Cognitive modelling of information retrieval is thus concentrated on human knowledge (user, reference worker, producer of knowledge) and systems knowledge (also human, but in a way organized, reflected, synthesized and recorded multimedially, sometimes referred to as task knowledge). As for the new cognitive paradigm in information retrieval, we could distinguish cognitive abilities, cognitive processes (general and specific), and cognitive styles. (Allen,B.L. 1991). Apart from these, according to Dilthey, each individual has his own comprehensive worldview (lived experience) which influences his ability to understand and act meaningfully in the world. This is a part of the context and the environment in information retrieval.

User studies in information science revealed the important roles of the task knowledge (specific goals of user's information-seeking behaviour) and the domain knowledge (knowledge of the topic being searched, i.e. the general subject area). However, we would like to include in our modelling especially the individual cognitive and creative context which is really difficult to define in terms of traditional methods of information retrieval research.

Cognitive processes can be determined as mental activities such as thinking, imagining, remembering, and problem solving. General cognitive processes include the dynamic human thinking and occur in the form of cognitive behaviour. Specific cognitive processes may be determined by learning, memory (remembering), comprehension. Cognitive abilities are in fact individualized combinations of cognitive processes within each individual which reflect his preparedness to act with the use of them. Cognitive styles are developed out of the use of cognitive abilities and are stable preferences in the ways people think, learn and solve problems.

These methodological issues should be clear in developing new fundamentals of an information retrieval course. Once the objects of modelling are defined we can start to identify their categories exploring their characteristics, attributes and dimensions. Then the connections between them should be found and relationships identified. This modelling should cover one or two topics within the information retrieval course. Each student should look for the meanings through creating them within the course - that is the way students could understand the pocess of cognitive modelling of information retrieval.

The outline of cognitive modelling of information retrieval has been backed by the principles of the object and managing systems models which take into the account both context and internal structure of knowledge. The two issues can shape and direct the behaviour of users in information retrieval and the process is represented by the form of a spiral. As the activities in information retrieval are evolving, knowledge organization follows the representational flexibility within a certain situation. That is why it is not practicable to seek for the universal knowledge organization means in information retrieval. What is more valuable, however, is synergetic modelling of its part performed in the known situation.

5. The Interdisciplinary Model of Knowledge Structures in Information Retrieval

5.1 Outlining the principles
The outline of the principles shaping the model is demonstrated in Fig.1. The frame of cognitive and physical paradigms in the theory of information retrieval points to both cognitive modelling and dynamics of the process and to new multimedia information technologies as an indispensable part of information retrieval. Environmental synergy includes cognitive objects (users, information workers, producers of knowledge) and systems objects (information technologies, situation and context of information retrieval). These objects are linked in information retrieval through knowledge (as we have specified - domain and task knowledge, but also cognitive issues embodied in knowledge) and its main processes which are determined for information retrieval as organization and utilization. The dynamics of knowledge is represented by its communication reflecting the change and complexity caused by special objects of information retrieval. Thus, fig.1 should demonstrate the spiral of information retrieval with respect to knowledge organization. The central role of information retrieval can be viewed not only as a means of knowledge organization and utilisation, but also as a creative impetus for knowledge production, filtering and synthesis in the direction of the spiral movement of knowledge in its cycles.

5.2 The interdisciplinary structure of the model
The interdisciplinary structure of the model contains a number of contributing disciplines including cognitive sciences, computer sciences, psychology, mathematics, logic, information theory, communication, linguistics, systems theory, management, and

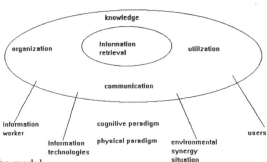

Fig.1: Outline of principles of the model

some more traditional in information science (librarianship, classification, information theory). From this viewpoint information retrieval theory has been traditionally split into two large directions: a) text retrieval and indexing, b) data and systems management. The further coexistence of these directions, however, (comprising so many disciplines) can be enabled in our model by the harmony of cognitive and physical paradigms, by their balance within the theory and methodology presented to our students.

The core disciplines entering the information retrieval course in our new concept represent cognitive science, natural language processing, artificial intelligence, learning theory, knowledge representation, problem solving and planning. A survey at library schools (Kranch 1992) has shown a lack of these disciplines in subjects dealing with knowledge organization at the master's level of studies. Because of the interdisciplinary structure of our model we propose the inclusion of the fundamentals of these disciplines into the curricula in order to develop a real understanding of information retrieval functions in knowledge organization.

5.3 Attributes and constraints

The attributes of the information retrieval objects viewed from the knowledge perspective form the complexity of non-linear principles as a rule of information retrieval. They should be specified because they cause the results of cognitive processing through perception, learning and behaviour of subjects in information retrieval. In fact, we should speak of multiple models pertaining to special subjects of information retrieval developed of creative ordering principles for knowledge processing by users. The constraints are therefore represented by modelling into the language order, as language structure brings about its own laws. The problem is then in the process, transformation and transition of knowledge in the interaction of the subject with the environment of information retrieval that could result in triggering new order structures during the retrieval processes. That is why the constraints should be described as the characteristics, ways, and the extent of language metamorphosis.

Following these constraints the students should be shown at least such attributes of knowledge objects in information retrieval as the form, the source of information, content complexity, context, intellectual integrity, mode of recording. The order structures of knowledge organization in information retrieval are based on the organization of concepts which form the important means of cognitive self-organization and reflection. Concepts as the attributes of knowledge objects should be within the information retrieval course elaborated in all their aspects important for evolution and preservation of environment and life including material and energetic, perceptual, rational, emotional, formal, intentional, and value-driven (ethical).

The outlined attributes and constraints are in information retrieval the starting-points of representing and organising dynamics and complexity of knowledge in the way of a number of corresponding layers mutually connected in networks of simple structural patterns. This way of explaining information retrieval, in my opinion, is more correct than the traditional ones, in that it approaches the truth connected with presentation of new research results.

5.4 Deep Networks of Processes

Based on our meta-conceptual model and the grounded theory approach (Ellis 1993) to information-seeking patterns of researchers we derive a model approach to the determination of processes in our model. The processes include starting, chaining, browsing, differentiating, monitoring, extracting, analysing, verifying, ending, conceptual ordering, synergetic synthesizing. In another network they may form special combinations for special purpose within information retrieval including understanding, interpretation, description, reproduction, translation, explanation and expression. The networks in information retrieval are concentrated around each of these (and other additional) processes and linked together in deep structural networks. Following the iterative process of the "hermeneutic spiral" it is important to differentiate preunderstanding (self-reflection and reflection of environment including previous knowledge, experience, training, interpretation) and understanding as the process oriented towards knowledge in information retrieval. The change of preunderstanding into understanding is present in information retrieval in the spiral movement of knowledge.

This model tries to take into account the active role of the individual user in a multiplicity of specific models which can be formed by the single user and by various users during information retrieval. The boundaries of the models may be further specified by the mentioned processes connected in networks. It enables us to allow for and involve the individual interpretation and the personal point of view even within the modelled information retrieval and its environment.

Advances in philosophy, social sciences and sciences can be no longer ignored. Information retrieval theory needs elaborating their research results from its own point of view. One of the attempts could be the presented outline of our interdisciplinary model which, of course, needs further deepening within the specialised courses linking together knowledge organization with information retrieval.

6. The Practical Viewpoint (Information Retrieval, Information Consolidation)

Practical implications of the new approach to information retrieval can be seen in information systems design, with particular regard to information (reference) services modelling and planning. The service emphasizes the user within the interaction of cognitive paradigm and physical paradigm (and any other new paradigm which may occur in knowledge organization and utilisation). The concept of virtual reality library proves that we should think in new ways of knowledge organization and utilisation in information retrieval unless we want to let information retrieval research go to other disciplines. What will then be the role of information workers or librarians? What knowledge and skills should the graduates have in information and library science? These questions are very important even within the social development in central European countries.

The special information service model can be elaborated within the courses following the principles outlined in the interdisciplinary model. It should take into account the special environment of the single library and information centre with the objective of finding new ways of knowledge organization including environmental knowledge and information. Thus, we should strengthen the methodological issues of information retrieval courses as a response to the urgent need of a new quality of education.

Based on the previous sections, I would like to present the outline of the core program for elaborating the courses in Information Retrieval and Information Consolidation within our curricula with the new concept of knowledge organization. The mentioned theoretical foundations are here applied to the practice of the course teaching. However, the detailed elaboration of the course core is then specifically applied to each of the two courses.

Outline of the Course Core

Objectives:
- creative reflection of information retrieval and its environment using confirming, new, pragmatic information
- understanding information retrieval in terms of the interdisciplinary model, its environment, meaning and purpose in information processing and information science
- building an own opinion based on the presented concepts and philosophical background
- cultivation of professional and academic qualities of reflective thinking, critical and evaluative analysis and effective communication, identification and resolution of problems, interpretation and evaluation of knowledge

Additional specific goals:
- each student should be able to demonstrate his abilities and skills in the solution of an information retrieval problem
- development of oral and written communication and original research work abilities and skills
- evaluation of other students works using given and own criteria
- survey of the developmental trends and the history of information retrieval, conceptual structure (terminology) and principal professional literature

Main topics to be included:
1. Fundamental context of knowledge organization in society. Philosophical streams and knowledge organization. Terminology. Theory, methodology and application. Professional literature.

2. Main concepts of information retrieval. Transition to the contents of knowledge. Relationship of information retrieval system to information system. Cognitive paradigm, physical paradigm. Philosophy of information retrieval. Information retrieval chaos and harmony.

3. Contents, position, functions of information retrieval. Modelling information retrieval. Cognitive modelling. Fundamental interdisciplinary model of knowledge structure in information retrieval.

4. Historical perspective (philosophical development). Basic stages. Empirical approaches, theoretical approaches, experimenting, testing information languages. Information technologies and techniques in retrieval. Building retrieval strategies and tactics.

5. Derivation of specialized information (reference) service models. Elaborating on the knowledge structure model, determination of deep networks of processes. Derivation of rules and laws of information retrieval.

6. User modelling in information retrieval. Cognitive modelling applications. Recall, relevance, precision, pertinence - forming new evaluation criteria. Knowledge-based and expert systems in information retrieval.

7. Information retrieval management. Natural language processing - philosophy, development, applications. Text and knowledge management. Multimedia and virtual reality, the position of information retrieval.

8. Interactive, iterative cycling and spiral dynamic movement of information retrieval. New methods and techniques of knowledge representation. User studies. Qualitative research methods. Quantitative research methodology.

9. Information retrieval environment. Current changes. System of values, ethical considerations. Creative user and personal models of information retrieval.

10. Future. Advanced developmental tendencies. Examples of application in library and information systems. New models of services.

6. Experience in Teaching

Our students do not like the theory courses. They are too practical (or lazy) to be interested in deep theoretical issues of information processing and information retrieval. On the one hand, it is not surprising within the social and economical context in our country. If one wants to be successful, one needs money which is not to be made by theoretical research work. On the other hand, the courses on Information Technologies or Hypertexts should not teach students how to treat special software packages, because these skills can be taught within other educational facilities. The university education should preserve its academic characteristics. In the Information Retrieval Course it is possible through the development of "knowledge skills" of students in finding new ways of processing changing environmental influences on knowledge in information retrieval, i.e. comprehension, analysis, synthesis, evaluation, creativity, application, etc. (to name just a few of the large scale of skills taught at university level). This requires a lot of abstraction and generalization and the development of creative theories.

That is why one of the tasks of the course is to reveal the need for theory, and, what is more, the new system of values brought to information retrieval theory and practice from its societal environment and from the new paradigm should concentrate on the intensional and inherent development of knowledge based on new cognitive qualities, qualitative methodological approach using new technology as a tool for special objectives of knowledge processing and use.

In the future, professional workers in information retrieval should be able to manage new methods of knowledge analysis and organization, as well as to model users behaviour and their information needs in the information retrieval environment. That is why our course emphasizes especially interpretation and evaluation of knowledge. This will become more and more important both generally and applied to various areas of human activity including environmental knowledge organization as it is so close to the set of principal environmental values of man in this world. Thus, if information retrieval reflects new philosophical paradigms it develops its own information retrieval philosophy which could enrich even some of the streams of general philosophical reflection.

7. Conclusion

Contrary to normalization and mathematization of information retrieval, we have outlined the principles of a new approach to teaching and research of information retrieval and consolidation applying knowledge organization issues. As a potential of knowledge organization, fundamental concepts were described in possible cognitive modelling and as a model of knowledge structure in information retrieval. Considering knowledge structures new techniques of information services emerge. This could serve as an interesting example of current research and practice applied to teaching, especially from the methodological perspective. It is important to strengthen the transition from simple data and information to deeper knowledge and its creative interpretation, analysis and evaluation. This transition is reflected in the usefulness of the retrieved and processed knowledge for special types of user models managing knowledge chaos in information systems. It calls for a serious break in traditional thinking of library and information specialists in Slovakia, but also in other countries. The future societal development will probably emphasize the role of information retrieval in knowledge organization covering all sectors of our society. That is why our professionals should be prepared to seek for new ways of knowledge organization and utilisation, evaluation and interpretation. The area of information retrieval will then be affected by more and more new disciplines and it should respond by its own contribution to the shared problem of knowledge in society. We should recognize its creative issues and special ways of addition of new knowledge. However, we must move forward a little bit from the

178

deterministic and linear approach admitting the complexity and dynamics of knowledge organization in our models.

The challenges of a future information retrieval development are mainly caused by structural changes brought about by new information technologies that contributed to non-linear rethinking of the problem. Librarians of the future will then definitely need more creativity, communication and problem-solving skills. Apart from understanding management techniques and skills, they will have to understand cognitive processes in knowledge organization and utilization. Self-reflexive creativity and a personal point of view will no longer be a weakness, but a new starting-point for information retrieval modelling covering the whole range of knowledge organization issues.

References

(1)   Allen, B.L.: Cognitive Research in Information Science: Implications for Design. In: Annual Review of Information Science and Technology. 26(1991)p.3-27

(2)   Bradley,J.: Methodological issues and practices in qualitative research. Libr.Quart. 63(1993)No.4, p.431-449

(3)   Ellis, D.: Modeling the information-seeking patterns of academic researchers: a grounded theory approach. Libr.Quart. 63 (1993) No.4, p.469-456

(4)   Ellis, D.: Paradigms in information retrieval research. J.Doc. 48 (1992) No.1, p.45-64

(5)   Jaenecke,P.: To What End Knowledge Organization? Knowl.Org. 21 (1994) No.1, p.24-28

(6)   Jeng, L.H.: From Cataloguing to Organization of Information: A Paradigm for the Core Curriculum. J.Educ.forLibr.and Inf.Sci.(JELIS) 34(1993) No.2, p.113-124

(7)   Kranch,D.A.: Teaching Artificial Intelligence and Expert Systems: Concepts in Library Curricula. J.Educ.for Libr.and Inf.Sci.(JELIS) 33 (1992) No.1, p.18-34

(8)   Löckenhoff, H.: Systems Modeling for Classification: The Quest for Self-organization. Knowl.Org. 21(1994) No.1, p.12-23.

(9)   Paisley,W.: Knowledge Utilization: The Role of New Communication Technologies. J.Amer.Soc.Inf.Sci. 44 (1993) No.4, p.222-234.

(10) Steinerová,J.: Problems of knowledge representation in information system transformation. Dissertation thesis. Bratislava: FFUK 1992. 169p. - (in Slovak)

(11) Sutton,B.: The rationale for qualitative research: A review of principles and theoretical foundations. Libr.Quart. 63(1993) No.4, p.411-430

(12) Zelger,J.: A Dialogic Networking Approach to Information Retrieval. Knowl.Org. 21 (1994) No.1, p.24-28.

Werner Pillmann
International Society for Environmental Protection, Vienna

Experiences with Environmental Networking within CEDAR

Abstract

The "Central European Environmental Data Request Facility" CEDAR is a project of the Austrian Federal Ministry for Environment, Youth and Family in the International Society for Environmental Protection in Vienna. This data center is to provide environment-related information and to serve as a clearinghouse for any relevant requests. With the help of electronic communication, contacts were established to the countries in East-Middle-Europe and the international environmental protection agencies. In the first part of the program, existing since 1991, a data-collection by questionnaire was undertaken. Some 700 Experts working in environment-related fields in universities and administration indicated their interest to collaborate with CEDAR.

In order to comply with the functions of a Clearinghouse and to offer environment related information within given limits concerning personell and organization, a computer (workstation) was installed which is connected to Internet via the computer center of the University of Vienna. Further retrieval possibilities exist from commercial database services, such as Datastar, ECHO, and the host of USA/EPA (Environmental Protection Agency).

The databases offered by CEDAR comprise such areas as: environmental technologies, environmental activities, experts, training and further education in the areas of agriculture, environmental literature, innovative technologies in sanitation of cotaminated sites, and environmental monitoring.
To facilitate access according to an information request a "Gopher" was installed at the CEDAR-Workstation. This contains information - in some cases also multilingual information - such as:

CEDAR Information (Expert-Database, Environment
    relevant Information)
United Nations Environmental Programme: Infoterra
Expert Database
Central & Eastern European Libraries Database
Regional Environmental Center for Central and
    Eastern Europe
Wold Wildlife Fund U.S. (WWF)
Environmental Gophers (Austria, IIASA, USA/EPA,
    Envirolink, Ecogopher etc.)
UN Gopher servers (WHO, UNDP, UNICEF etc.)

For the purposes of networking, so far three workshops on the topic "Working with Internet" were held. The participants came from the Infoterra National Focal Points as well as from the East-Middle-European area. For the collaborators of the Polish Institute for Environmental Protection in Warsaw a remote-network-training was arranged from the Vienna base. At present two subscription-lists are taken care of. One of the lists serves exclusively the information exchange of the Programme Activity Center of UNEP, the Infoterra National Focal Points, and the

Regional Service Center. The other one serves as a discussion forum for the general public and is - at present- accessed by some 1000 persons per week. Closer connections exist to US-EPA and WWF USA.

In the "Memorandum of Understanding" of 1993 CEDAR was designated as a "Regional Service Center" of the UNEP/Infoterra-Program for the Middle-European area. In the first CEDAR/Infoterra Newsletter eight East-European countries described their activities as "National Infoterra Focal Points"

CEDAR is accessible via

The Gopher Host: pan.cedar.univie.ac.at PORT: 70
and via e-mail: cedar-info@pan.cedar.univie.ac.at.

Vladimir Kašša
NCI, Bratislava, Slovakia

Networking of Environmental Information Libraries, Information Centers in Slovakia

Abstract: A communication infrastructure for networking in the R&D academic community in Slovakia is presented. The short historical overview of important milestones is followed by a description of the present situation. Available networks, which offer worldwide connectivity and various kinds of services, are described, analyzed and evaluated from the users point of view. The main part of this paper is devoted to those problems confronting network users and network providers caused by the present, continuously changing work-environment. These problems need to be solved to enable users to concentrate on their tasks and to reach their goals whithout unnecessary delay.

1. The history

It was rather optimism than real need for a new telecommunication infrastructure that led us to start computer network activities in Slovakia more then 20 years ago. This activity took place almost exclusively in the academic area (R&D Institutions and Universities) and has resulted in the development and operation of the datagram network EPOS in 1984. At the end of the eighties the transition to the OSI network architecture was planned and in 1991 successfully realized. Since this time, an X.25 network named UAKnet is in operation. However, already during this transition period and shortly after, the necessity and urgency of a multiprotocol communication infrastructure was recognized. Because of such need, the telecommunication infrastructure had to be and was upgraded to a modern one that was able to fulfill the needs of that time, all present needs and many of the expected future needs. Slovak Academic Network SANET supporting the TCP/IP suite of protocols was designed and put in operation early in 1992. Besides these activities in the academic area, Slovak PTT has started its own activities in the area of public network services and since the end of 1991 the Slovak public data network EuroTel is in operation.

Networking milestones in Slovakia:
- ▸ October 1980 — start of networking activities (10 years after US four-node ARPA went 'on the air' and five years after French CYCLADES)
- ▸ September 1982 — experimental three-node IAC network went 'on the air' at the '82 International Computer Networks Conference
- ▸ October 1982 — connection to Eunet for E-mail services
- ▸ November 1984 — EPOS connected with 17 Institutions went into operation
- ▸ May 1986 — IAC obtained the license for operating its private computer network EPOS (but only ON-line Terminal - Host access services have been allowed. All other services, e.g. File transfer, Electronic mail, remote resources sharing, as well as any transborder communication has been forbidden)
- ▸ All year 1987 — Golden Age of EPOS (67 Institutions connected and volume of data transfer increased by about 70% a year)

▶	End of 1989	- Political and economical changes behind 'The iron curtain,' had a positive, and unfortunately also a negative influence not only on our networking activity.
▶	All year 1990	- Transient period (old EPOS was to be retired and the new X.25 network UAKnet was to go in operation)
▶	January 1991	- UAKnet was fully operational offering inland X.25 connectivity (almost one year before the start of the Slovak PDN EuroTel)
▶	July 1991	- UAKnet international connection via Austrian Academic Network ACONET (IXI and DATEX-p)
▶	July 1991	- Upgrade of the Eunet E-mail connection to Amsterdam from dial-in to a leased line (via ACONET)
▶	November 1991	- Slovak PDN EuroTel started its operation
▶	January 1992	- Slovak Academic Network started its operation
▶	September 1992	- UAKnet connection to EuroTel
▶	Since Jan. 1993	- Multiprotocol communication infrastructure is available including Gateway services between UAKnet, SANET, INTERNET and worldwide PDN's

2. Present situation

Except for some preparatory work, the main work concerning the transition to the new telecommunications infrastructure for the academic area began in 1991. The major past and present activities concerning the creation of the new network infrastructure deal with the following seven problem areas:

◆	LAN	- local area networks of particular institutions
◆	WAN	- wide area multiprotocol backbone networks
◆	LAN-WAN	- network interconnections in multiprotocol environments
◆	ACOnet CESnet EuropaNET	- cross border connectivity in multiprotocol environments
◆	SANET	- creation of a modern, powerful communication infrastructure for all academic communities (R&D and universities) in Slovakia
◆	INTERNET	- connectivity to world wide information sources offering their information almost free of charge
◆	EuroTel	- connectivity to domestic and world wide information sources offering their information almost unrealistically high priced

The present networking situation in Slovakia that resulted from the activities mentioned above provides inland and world wide connectivity to academic and commercial users. However, different protocols and tariffs have to be taken into account. Both, academic and commercial users are burdened heavily by questions of protocols and tariffs.

While there is little impact from TCP/IP on commercial networking, there is a significant commitment to proprietary standards. The opposite is true in academic networking. Because of present EuroTel's tariffs (EuroTel is the third most expensive PDN in Europe) it is unrealistic to move the academic network traffic that is already heavy and explosively increasing by a factor of four each year to the much more expensive EuroTel network. This is not crucial to our commercial users because of their presently low traffic. What is very important is that the current needs of all users are met and that future plans and investments aim towards a more 'open' approach.

Thus we are led to leave it to the user rather than to the expert to make the decision on which network and protocol to adopt and use. We are sure, that the present communications infrastructure is giving them this possibility.

3. Problem's presentation

Even if many domestic and foreign institutions being involved, it is necessary to underline, that the hard core of the work has been and will be carried out by the limited number of individual networkers. Appreciable help from abroad and also the possibility of the closer and more efficient international cooperation has to be acknowledged as a positive result of political changes behind 'The iron curtain.' But, as every coin has two sides, the accompanying economical changes have also made our lives harder; sometimes much more than we expected. The cutting in R&D programs in our country were the logical consequence. This in turn resulted,logically, to a decreased demand for all kind of information.

However, things are going from bad to worse. R&D institutions like our Institute for Applied Cybernetics (IAC) now renamed the National Centre for Informatics (NCI) are dying at a weekly rate. It is getting increasingly harder for them to survive. R&D programs at universities have higher hopes to survive. Unfortunately, the people involved in such programs are either early beginners or their R&D programs are ill. They are namely living with the naive idea that, electronic mail, electronic news and the collection of public domain software could fully satisfy their needs for carrying out their R&D programs. But all of us who are involved in and responsible for the creation of a modern telecommunications infrastructure strongly believe that such infrastructure will sooner or later be a necessity also in our country. Such believe lead us to and resulted in the present telecommunications infrastructure not only for academia but also for all Slovak users.

After more than 15 years of activities in the data communication and in networking area I recognize, especially in our country, three disparate worlds in which people are living alongside each other but not together. These worlds are:
- PTT people,
- Private network provider and
- Network users.

These worlds are living alongside each other having very different views on data communications problems and little desire or ability to understand each other.

Rules governing the PTT world:
* Every bit and every kilometer costs real money (who uses network has to pay for it).
* Circuit switching is superior (even thinking about something else is not allowed)
* The network is independent of its attached devices (it is always possible to point to the attachment - connector).

Rules applying to the private network world:
* Bits, bandwidth and distance are cheap, the available technology determines the costs.
* Packet switching is the rule (why go through the trouble of opening a circuit for a single transaction?).
* Devices are considered part of the network (the user interface is more than a connector).

What users want:

* Network infrastructure should be priced realistically (tariffs should not be large random numbers).
* The average user does not want much more than what he has (but it should work properly).
* The average user is ready to accept a certain amount of dependence of provided services on the distance between communication partners.
* Because of tremendous differences between user profiles, the definition of the average user should be treated very carefully (in real life there are user groups).

Now we turn our attention to the present, urgent problems.

* No. 1: Exponential growth of network traffic. Volume of data, being transferred in our academic network, is increasing by a factor of 4.4 every year. Factors for individual lines vary between three and 10. Many people treat this fact as a positive sign of network development. But I think, such optimism is very questionable. Such evaluation may be valid only during the very early stages of network development and that is not our case. We have not a lot of time left to take control over such uncontrolled growth of traffic. If we fail to solve this problem soon and leave to be solved by itself, we must be ready to face many, much more serious problems in the future. Every two years upgrade of line speed by the factor of 10 could be (very soon) not only a question of money, but of available technology too.

* No. 2: Exponential increase of tariffs. Twenty years ago we paid for a leased line 380 crowns. Now, for the same line we are paying almost 20 times as much. The traffic over our PDN EuroTel is the 500th part of that carried carried by the private UAKnet, but it costs us twice as much as on UAKnet. Is such tariff structure realistic or should every institution have its own WAN network?

* No. 3: Commercialization of the academic network. The Slovak academic network SANET as part of INTERNET is going to change its user policy toward commercial users. This sounds very nice, but it is necessary to consider the state of things in our commercial area. Our commercial companies are not so well established as those in other parts of the INTERNET world. To apply the same, common commercial user policy in both areas would be very short-sighted. The key role in our commercial companies is played by those people, often the best, who proved their courage or were forced to leave our academic area. Such short-sighted commercialization could cause them loss of the connectivity which they already had and used. This could be a disaster for both and of course for society as well. Because I do not see a more effective and more reliable source for funding of R&D and university programs than the state budget, a budget filled with money coming from healthy and profit making companies.

* No. 4: Security and privacy in the academic network. SANET as part of INTERNET belongs to a global system of interconnected networks that share a single address space. It is used for electronic communication, data retrieval and computer access. Security and privacy are a key concern, in particular, to commercial users. Once a company is connected to the network, it is connected to virtually to all participants in net. The only perfectly secure and private network is one that is not connected to the outside world. Present E-mail message traveling through our networks is the

electronic equivalent to a note written in erasable pencil on the back of a postcard and then putting it in the mail. Password protected access to computer looks better, but it is necessary to consider its 'snowball' effect. The more passwords you have, the more passwords will be broken.

- No. 5: Domestic versus international traffic. Statistics concerning our domestic versus international traffic are alarming. Two 64 kbps international lines are carrying four times the data that all our domestic carry lines together. International traffic increases by a factor of 5.1 each year, while domestic traffic increases only by 3.2. Worse yet, almost all data crossing the border are coming from abroad. Only a negligible part of the data is of Slovak origin.

4. Proposed solutions

To propose an efficient and successful solution of all the above-mentioned problems is always easier said than done. The place to start is a comprehensive cause analysis that takes the complexity of the problems and their mutual dependence into account. Prioritization of the list of causes that have to be dealt with is the only way to a solvution to or to a bypass of our problems. This could be tedious and time consuming, but there is no shortcut. Simultaneously, it is desirable to make only incremental changes and to carefully evaluate the resulting gradient.

Which is the first step to take? I suggest opening our networks much more toward the 'real life' users. The present user group contains almost exclusively the 'networkers' themselves; that half percent of peoples having access to computer networks are using 95% of these resources. They are able to bring every network down on its knees. But real life users do not want much more than they already have. It should only work properly. The majority of such network users can not cause 'astronomical' increase in traffic. The right question is: How to manage it? In my opinion, the answer surprisingly easy. We have succeeded to network our computers, but totally failed in the 'networking of people.' Networking them around their tasks, jobs and businesses. That is the challenge to our leaders, coordinators and managers to show and to ask their people to work more efficiently by being 'doubly networked.'

What to do as the next step? Next, in my opinion are the tariffs. The present tariff structure is a serious barrier to cooperations. At first glanc all tariffs seem to be a mostly large random numbers. But a closer look reveals a very strong inverse relationship to the degree of deregulation and demonopolization of PTT's in the country of its origin. The country (USA) with the highest degree of deregulation and demonopolization has the lowest tariffs. Vice versa, the state owned and monopolized PTTs (Slovakia) belong to the most expensive. This is of course no surprise. Only competitive markets are able to set tariffs realistically, and that is what the 'real life' user needs. If this is true, and I am sure it is, it is not an unsolvable problem to bring order into our sky high tariffs. That is a challenge to our law makers to speed up our PTT's deregulation and demonopolization process. There are many users, especially in the academic area, but not only there, who would like to have zero tariffs for datacommunication and not only for this service. This is understandable in the present transient period of our society. A lot of information is widely available without the need to pay for it. In such circumstances every communications tariff seems unrealistically high to the user. Very often, even modest communications costs keep people from obtain the desired information. It would be very nice and not at all unrealistic, to give to all people equal right and access to retrieve every information that is free of charge without having to pay for the delivery from the source to the place of destination. That is a further challenge, in this case to our governmental Commission on Informatization, to fund such infrastructures. Is it taking its own Society Informatization Program seriously?

What about commercialization? Our academic network SANET as a part of INTERNET suffers from two fatal flaws that make it unsuitable for anything more than trivial traffic. One is its poor reliability; the other its total lack of security. Both problems stem from the very nature of the SANET, which means they can not be solved without transforming SANET into a totally different kind of network.To make matters worse, there is no one to blame for the problem since no one is responsible for what happens at SANET. When and if such transformation will happen is a question that is very difficult to answer now, but the commercial companies on SANET can contribute to and also be forced to support such a transformation. The first step (payed vacations) has been achieved and the next steps (reliability, security and privacy improvement as well as increased capacity of particular access points) should follow soon.

What to do for security and privacy improvement? One of the most common approaches to security taken by commercial companies is to create a firewall using a pair of routers with a host between them. The router on the SANET side lets the host communicate freely with the SANET, and vice versa. The other router is set up to prevent unauthorized traffic from reaching the company network. Another way is to create a virtual network using tunneling technology. Such virtual or pseudo network can add powerful and useful security control. In all cases SANET as access provider should come up with multilayered security schemes that make it possible to opt for different degrees of protection. That is the challenge for the SANET people and other network providers.

How can we change the present unhealthy traffic composition and its unhealthy direction of change? That is the most crucial and difficult problem. It can take not just months but years until our people recognize, that they themselves must solve their problems. They must cooperate and communicate, of course and first of all, in their own country or community. They have to recognize that mere information pumping cannot persist for ever. That is a challenge to all our people.

Conclusion:

We started with optimism on the road to networking in our country. Now, I believe, we are still pretty away from being fully justified in our optimistm, particularly in the networking area. No wonder, we have not always been completely. But in spite of this, we ca proudly look on a great deal that has been achieved, particularly in R&D and academic networking. We were always trying to be on the look out for change, not only for needed change, but also the opportunities for change. Unfortunately, it is not optimism but pessimism that is governing us now. Pessimism is one of our great problems not only in telecommunications technology. However, we must strive for optimism, or at least for relative optimism. Such relative optimism requires more than someone who uses informatics to provide better solutions to complex problems. Optimism is also needed for the solution of organizational and political problems. Many of these problems are forced on us by the simple fact of our people's inertia, especially of users and experts in telecommunications technology.

Eva Stengård
Swedish Museum of Natural History, Stockholm, Sweden

Production of Nordic Standard Environmental Term Lists

Abstract: *Nordic Code Centre* (NCC) was a project within the *Nordic Council of Ministers*. Through 1985-1993 NCC developed and maintained several term lists on biological species and other terms relevant for describing environmental parameters. NCC's steering committee consisted of one representative from each of the participating countries. The responsibilities for maintenance of standard term lists was distributed among institutions in the member countries depending on type of terms. Beginning in 1994 new term lists is to be produced in selffinanced projects. The former steering committee has been given the opportunity to work out directives for a new organization.

1. Nordic Council of Ministers

The *Nordic Council of Ministers* (see Fig.1) was founded in 1971 as an organization for co-operation between the Nordic governments. It has executing responsibilities for work done within the *Nordic Council* which aims at promoting cooperation between the Nordic parliaments and goverments. The *Nordic Council* consists of elected members whereas the Council of Ministers consists of ministers or other officials from the participating governments. Depending on the matters to be discussed, the Council of Ministers meets in different compositions (1).

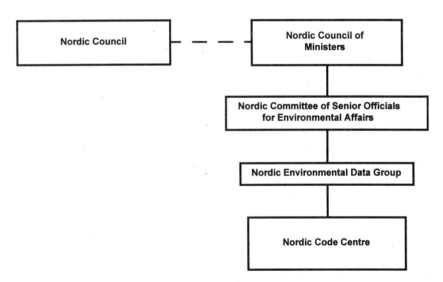

Fig. 1. Organization of Nordic Council of Ministers illustrating the position of the project NCC. The connection between the Nordic Council of Ministers and the Nordic Council is also shown.

Environmental cooporation is conducted by the *Nordic Committee of Senior Officials for Environmental Affairs*. It is aimed at contributing to the improvement of the environment and forestall problems in the Nordic countries as well as internationally (1). The committee has working groups engaged in different aspects of environmental affairs. One of its former working groups was the *Nordic Environmental Data Group* that was the initiator and financer of the project Nordic Code Centre.

2. Nordic Code Centre

Nordic Code Centre (NCC) was started in 1985 to expand part of the work done by a Swedish environmental project to the benefit of all Nordic countries. The Swedish project aimed at establishing routines for biological inventories and among other things it recognized the necessity of using standardized data. The project NCC was concluded in 1993 when the *Nordic Environmental Data Group* was fused with the *Nordic Environmental Monitoring Group* to form the new *Nordic Environmental Monitoring and Data Group*.

NCC developed and maintained several term lists on biological species, chemical/ physical determinands and other terms relevant for describing environmental parameters. The project had a steering committee consisting of one representative from each of the participating countries, appointed by authorities involved in national environmental protection.

The amount of contribution differs among the participating countries. Three of them had the managing responsibility for terms in an area of concern, usually in the form of one or several basic registers. Thus was the *Forest and Nature Agency* in Denmark, in cooporation with the *Swedish Environmental Protection Agency*, managing institution for habitats[1] and other similar biological parameters. The *Environmental Data Centre* in Finland managed chemical and physical determinands and finally the responsibility for species was carried out by Sweden through the *Swedish Museum of Natural History*. The museum was also used as a secretariat for the project (3).

3. NCC Term Lists

The purpose of the term lists is to standardize environmental terms used among the Nordic countries and to exercise vocabulary control on terms within environmental data systems. This is necessary both to ensure information retrieval and to enable transfer of data in space and time. The value of reliable inventory data previously recorded from an area to use as reference data cannot be overestimated.

The theusaurus-like composition of the term lists provides a way to handle synonyms and groups of related terms. A sorting number enables sorting of terms according to the used classification. This is especially useful in term lists on species. Species term lists is based on the scientific names of species (Latin) and other lists is based on the English expressions for the included terms. All valid terms is provided with a unique short-hand representation known as a code which is mnemonic to its nature (see Fig.2). In early days when computer memory space was an issue the term lists was known as Code Lists.

Species terms are divided in a series of basic registers reflecting wellknown taxonomic[2] groups. All term lists are available as computer files and some are also available as printed publications. The basic registers for the term lists are continuously updated.

Despite the presens of systematic classification, the term lists are not systematic publications which is very important to point out. The systematics in a term list is by necessity a compromise.

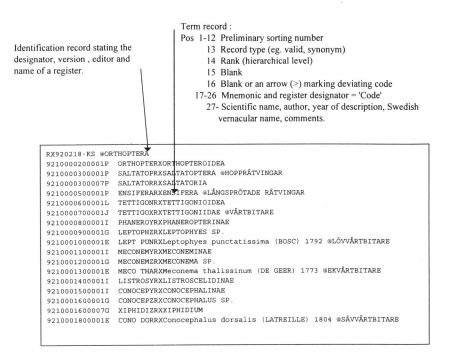

Identification record stating the designator, version , editor and name of a register.

Term record :
Pos 1-12 Preliminary sorting number
    13 Record type (eg. valid, synonym)
    14 Rank (hierarchical level)
    15 Blank
    16 Blank or an arrow (>) marking deviating code
17-26 Mnemonic and register designator = 'Code'
    27- Scientific name, author, year of description, Swedish vernacular name, comments.

```
RX920218-KS @ORTHOPTERA
9210000200001P  ORTHOPTERXORTHOPTEROIDEA
9210000300001P  SALTATOPRXSALTATOPTERA @HOPPRÄTVINGAR
9210000300007P  SALTATORRXSALTATORIA
9210000500001P  ENSIFERARXENSIFERA @LÅNGSPRÖTADE RÄTVINGAR
9210000600001L  TETTIGONRXTETTIGONIOIDEA
9210000700001J  TETTIGOXRXTETTIGONIIDAE @VÅRTBITARE
9210000800001I  PHANEROYRXPHANEROPTERINAE
9210000900001G  LEPTOPHZRXLEPTOPHYES SP.
9210001000001E  LEPT PUNRXLeptophyes punctatissima (BOSC) 1792 @LÖVVÅRTBITARE
9210001100001I  MECONEMYRXMECONEMINAE
9210001200001G  MECONEMZRXMECONEMA SP.
9210001300001E  MECO THARXMeconema thalissinum (DE GEER) 1773 @EKVÅRTBITARE
9210001400001I  LISTROSYRXLISTROSCELIDINAE
9210001500001I  CONOCEPYRXCONOCEPHALINAE
9210001600001G  CONOCEPZRXCONOCEPHALUS SP.
9210001600007G  XIPHIDIZRXXIPHIDIUM
9210001800001E  CONO DORRXConocephalus dorsalis (LATREILLE) 1804 @SÄVVÅRTBITARE
```

Fig. 2. Example of a species term list in its basic plain ASCII format

4. Classification Facets of Species

The remaining part of this communication will concentrate on species term lists as they form the larger part of the term lists produced by NCC.

A term list on biological species could of course be alphabetically arranged but that would not enable retrieval of groups that is used in different contexts. A classification of the species names in a list is therefore desirable. Classification may be done according to different principles. The traditional way to classify is according to their taxonomic position. Most literature on species uses this. Another way to classify is based on common characteristics. Related taxonomic groups have often several characteristics in common but most functional classification still overlaps taxonomical groups. Examples of functional classification is a classification according to ecological niche[3] or habitat.

5. Systematics

Systematics is the science dealing with the organization, history and evolution of life. It puts questions about how different life forms originated, how they diversified and how they have been dispersed over time and space. A discipline within systematics is taxonomy. Taxonomists describe and name life forms and arrange them in classifications reflecting patterns of relationship (see Fig.3). Taxonomy creates a language for systematics and other biological disciplines concerned with species (2).

The classification is done based on knowledge of the different organisms and as that knowledge grows the classifications changes. It is not uncommon for a species to be described several times under different names as the type organisms may be regarded as different species. The defining of a taxon and deciding what descriptions are synonymous is a matter of opinion by the individual taxonomist. Thus, a systematic classification of species is very dynamic and few traces can be found today of Linnaeus' *Systema Naturae...*

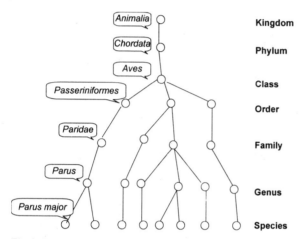

Fig. 3. Example of a systematic classification with the use of the principle hierarchical levels.

6. Species Classification in Environmental Protection

Work in environmental protection have use for several different types of classifications. Functional classification would be very useful but, as compiling the information needed to create them is very tedious and costly, systematic classifications are more commonly in use. An ideal classification in use by environmentalists should be stable and consists of all species names including synonyms (but only those) used in recognized keys[4].

Systematists recognize approximately 40 hierarchical levels, but only about a dozen of those are of interest for environmentalists. Observed organisms are in many cases not identified to species, rather to genus or family, and very seldom to a lower level than species. In environmental work the concept of interest is the taxon not the name.

7. Today - NCC Production of Term Lists

Even though the project NCC is concluded the managing institutions is carrying on the work according to the previous organization during 1994.

When producing a species term list, one or several editors are appointed by the steering committee to be responsible for compiling terms and preparing the draft version(s). In some cases the editor also prepares the version to be released but usually this version is prepared by the managing institution.

The editor contacts, with or without the help of the steering committee, one or more taxonomists to review names and classification compiled for the list. A list is primarily based on currently used keys in the Nordic countries and use a widespread systematical classification. The latest systematical classifications are avoided unless they have been accepted in a large circle.

Before release, the list is referred to the member countries for consideration by representatives of the intended users.

8. Today - Revisions of Species Term Lists

No term list is complete when it is released. For instance, there are always names and synonyms in use not anticipated by the editor and there is also a continous publication of new species. As previously mentioned the systematic position of a taxon quite often change. Therefore, a basic term register is constantly in need of updating.

NCC discriminate between two types of revisions, minor and major. A minor revision may be the inclusion of a few terms or a new classification of a few taxa that don't disturb the systematic numbering. Minor revisions are usually performed by the managing institutions upon request.

A major revision may be triggered by one of the following reasons. The term list is requested to enlarge the geographic coverage, large number of new taxa needs to be included or there have been a major revision of the systematics of the taxonomic group in question that has to be reflected in the list. A major revision usually requires a renumbering of the list and will essentially create a new term list. As with other new term lists the work is usually done by an appointed editor. All term registers are identified with a register designator and when a major revision has taken place the register for a taxonomic group will get a new designator.

9. Today - Distribution and Support

The national representatives in the steering committe are responsible for the contact with their national users. They have also been given the task of informing potential users of the existence and use of NCC Term Lists. They distribute files and publications upon request (3).

Most of the distribution is made by the managing institutions as they always have the latest version. Thus, the *Forest and Nature Agency* distributes the newly produced database on vegetation types, the *Environmental Data Centre* distributes the term list on chemical and physical determinands and the *Swedish Museum of Natural History* the term lists on species.

The managing institutions also answer any question of technical nature regarding the products they are managing. All products from NCC are free of charge but the national institution may take an administrative fee covering their expences in connection with the distribution.

10. Tomorrow - Projected Organization

The new *Nordic Environmental Monitoring and Data Group* (NMD) have recognized the value of the work done by NCC and have expressed their interest in a continued production of term lists. However they have not been willing to set up a new project. Instead, during 1994 they have given the former steering committee for NCC the opportunity firstly, to finish the work on term lists far in progress, and secondly, to work out directives for a new organization. At the time of writing those directives have only vaguely been discussed in the steering committe. However, some outlines are beginning to emerge.

A 'Nordic Environmental Term Group' should be formed with representatives from institutions or organizations interested in the continuation of term list production. Each representative should be financed by their own institution. NMD should also contract data hosts to ensure centrally managed registers. The term group should also attach technical consultants and topical experts on a more or less permanent basis. The technical consultants would be responsable for the training of editors or even function as such. The topical experts would have a responsability for the contents of the registers both at the primary compilation and at later revisions.

Term list productions will each form their own selffinanced projects.

11. Tomorrow - Term List Production

The production of a term list will only take place upon request and the requesting institution will be liable for all expences of a project, including a preproject. The task of the preproject is to evaluate the possibilities of creating a term list at a reasonable price and in a reasonable time. It can also check if there are more institutions interested so that the cost for each institution can be lowered. Customers do not have to be Nordic. If a decision is made on creating a new term list a project group is appointed with representatives from the term group, customer and all necessary consultants included. A project plan and budget should be ratified by both the term group and the financer(s).

All term list produced have to follow a common format but those customers requiring special formats should be able to get this in addition to the basic format.

There are a some problems that remain to be solved, among those is the question of who has the copyright on term list production

12. Tomorrow - Distribution and Support

Distribution and support is another area where there are several problems to solve. They are areas where some permanency is required and may have to be solved differently depending on type of term list and also depending on term list media (file or printed material). It is currently unclear whether the term group will have access to a secreteriat.

What is already clear is that the term lists will be available through Internet at an ftp server set up by the *Swedish Museum of Natural History*. Species term lists are already available through that channel. There are also discussions on setting up a distribution list for users to be able to subscribe on term lists through the networks and possible also a network conference for term list users.

References

(1) Nordic Council of Ministers: Nordic Concept for Environmental Data. Stockholm, 1993. 20p.

(2) Novacek, M.J.: The Meaning of Systematics and the Biodiversity Crisis. In: Eldredge, N. (Ed.): Systematics, Ecology, and the Biodiversity Crisis. Columbia University Press, New York. 1992. p.101-108.

(3) The Nordic Code Centre: NCC Coding System. Stockholm, 1990. 20p.

Notes

[1] The habitat is the local environment occupied by an organism.

[2] Taxon (pl. taxa) is any group of organisms, populations or taxa considered to be sufficiently distinct from other such groups to be treated as a separate unit. A taxonomic group of any rank, including all the subordinate groups.

[3] The concept of the space occupied by a species, which includes both the physical space as well as the functional role of the species.

[4] Devices used to identify specimens. Several types of keys exist, among those *natural keys* that follow a natural classification and *artificial keys*. The most convenient keys are arranged in dichotomous couplets.

Hartmut Keune, James McKenna
UNEP-HEM, Munich, Germany

National and International Aspects
of Environmental Information Management

Abstract: To exercise its mission of "...enhancing the compatibility and quality of information (enhancing harmonization) on the state of the environment worldwide...", HEM is developing a global environmental information system. "HEMIS" will coordinate information on who is doing what, where, how and why in relation to the collection, analysis and assessment of environmental information, thus making it a *meta-database*, or informational database of organizations, programmes, databases and monitoring activities including methodologies towards improved harmonization.

I. Introduction

An overabundance of data about the state of the environment is produced at different levels (global, regional, and local) at innumerable organisations worldwide. To make this data more useful and to avoid duplication, it is essential that data is collected in a compatible and comparable fashion. This is the major concept behind the term *harmonization* of data, which should not only mean harmonizing data *within* programmes, but *between* programmes as well (1).

In 1989, based on recommendations of the German National Environmental Advisors and the UNEP Governing Council resolutions, UNEP within the framework of the Environmental Assessment Sub-Programme (EAS), and the German Ministry of the Environment (BMU), jointly established a Centre in charge of Harmonization of Environmental Measurement (HEM): the UNEP-HEM office.

Since its inception, the HEM office has been involved in developing a framework for providing appropriate environmental information for the decision-making process. This requires meeting the demand for compatible and comparable data and information, reflecting the globalization of environmental issues, and the need to be able to measure progress on a range of issues which cross international frontiers. HEM's current activity areas include (2):

1. Harmonization of on-going and planned monitoring programmes in specific, selected environmental sectors.
2. Harmonization of classification systems in selected areas.
3. Compilation of inventories and surveys related to environmental measurement activities.
4. Coordination and liaison with other international agencies involved in harmonization.
5. Development of an information system to assist with harmonization - HEMIS.

The purpose of the planned Harmonization of Environmental Information System, HEMIS, (Activity Area 5 above) is to improve the state of, and access to, environmental information on a global scale. It will allow broad dissemination of information on who is doing what, where, how and why in environmental monitoring and assessment. The meta-data management system's principal objective is to offer its users current details of data sources (3). This should ensure that all relevant information is available when policy-makers propose public decisions.

HEMIS is to be a high-level information system. It will contain generalized information for planning and programme management, and for identifying the sources of more detailed data. It will not contain detailed primary data, but rather, information about data, and thus, HEMIS will be a "Meta-database". Since HEMIS will be housing "meta-data", it will be neither a duplication, nor a superset of already existing systems, but instead, it will serve as a compliment to them. Such a database can provide a tool for identifying areas in need of harmonization, and gaps in global environmental data collection.

It is hoped that within the framework of harmonization, three global observing systems (GCOS, GOOS, GTOS) will serve as the global structure for environmental monitoring and assessment. With this in mind, HEMIS is designed to serve three main user groups:

1. Decision-makers - especially those in developing countries who need access to harmonized information to make national decisions in a global context.
2. Research Organizations - especially those concerned with cross-sectoral interrelationships and modeling.
3. HEM's (and other) Harmonization processes.

II. User Requirements

The basis for user requirements are characterized by four key elements (4). The first element is the need for information exchange and communication. A clearly identified problem that has faced decision-makers and researchers in the past, has been the needless duplication of effort and waste of resources en route to their desired activities' objectives. Better channels of communication will reduce redundancy and permit users to benefit from the previous expertise of their colleagues. The second element targets a global user community. It is desired that all international users get involved actively by making themselves and their domain of knowledge known to HEMIS. One of HEMIS's main assets is that it will be extremely easy to access for developing countries, as well as developed countries.

A subject matter scope which uses a broad interpretation of "environment" is the third element. The key is an ecosystems approach, in which human activity and well-being are an integral part. Economic systems must harmonize with ecological systems and synergy must reign between environmental management and economic development (5). While we as a planet realize the implications that arise from the advancement of economics related to sustainable development, the maximum amount of information about ecosystems in the environment should be included in HEM's Information System.

The fourth element requires HEMIS to be a referral system or meta-database containing information about the main category areas of environmental information and programmes, environmental databases, environmental measurement techniques, models, reference materials, etc.

Concerning the organisational framework, HEMIS will aim to serve UNEP itself, especially in its commitments to environmental reporting and national capacity building, the rest of the UN system (particularly WHO, WMO, FAO, UNESCO, UNDP), International

Scientific Organisations and NGOs (like ICSU, IUCN, WRI), the international "Global Observing Systems" of GOOS, GCOS and especially the new Global Terrestrial Observing System (GTOS). HEMIS is intended to be the principal harmonization tool and meta-database to facilitate the planned network of terrestrial observation under this programme, government and international policy agencies and the general scientific community.

III. Design Objectives

A)   Content

Environmental data describe the status and trends in resources, the quality of the environment in addition to dealing with human activities. Through numbers, categories or texts, this data reflects environmental objects or phenomena related to occurrences in time and space (6). The initial scope of the data will emphasize international relevance of the entries, and will exclude site-specific data of only local significance. HEMIS will first focus on major international programmes. HEM's "Survey of Environmental Monitoring and Information Management Programmes" gives an idea of the intended geographic and organisational scope (4).

1.) Scope of Data and Organisational Activities.
a) environmental indicators
b) atmospheric ozone depletion
c) long range transport of air pollutants
d) behaviour and effects of persistent chemicals
e) behaviour of pollutants in soils and sediments
f) supply of reference materials and archiving of environmental samples
g) measurement of key parameters in global climate change
h) pollutants in natural waters and effects on natural resources
i) wildlife and protected areas
j) biodiversity
k) relationship between environment, human health and well-being

2.) Entities. The 3 main entities will include institutions, information management (data) activities (for example, programmes, projects, organisations) and product forms of the data (output such as databases, reports, software, etc.). From these, additional entities will be created to further define the data including experts (individuals), programmes, methods/models, classification systems, datasets, stations, addresses and communication.

3.) Data sources. Some data sources include HEM surveys, EEA's Catalogue of Data Sources, WCMC's meta-data, CIESIN's Catalog Services, Infoterra's expert information system, ESA's Directory of Space and Earth Science Data and GRID's datasets.

System design should facilitate direct input from related cooperating organisations (including inter alia, EEA, ESA, WCMC, CIESIN, GRID, MARC, WRI, World Data Centres, etc.) in an effort to make data more comparable and compatible. For this reason, interchange formats compatible with NASA's DIF structure should be used. (WCMC, EEA, HEM and GRID have already developed compatible formats.)

In order to have truly meaningful, complete, self-explanatory data that can be used by outside sources, it is imperative to include "circumstantial information" about the data, such as, what is being measured, how is it measured, by whom, when, why and where was it measured (6). In this fashion, the documentation in the meta-database will have concrete, lasting meaning. This will aid harmonization within and across sectors by making all information transparent and easily traceable or translatable by users unfamiliar with the specific data in question.

B)   Logical Access

To gain access to the system, on-line connection should be possible. For users without on-line capabilities, it will be possible to make requests.for portions of the meta-data that will be contained on HEMIS. Requests by fax, Email, postal service and telephone will be directed toward the appropriate HEMIS focal point who will distribute self-contained "stand-alone" software on diskette or CD-ROM.

By connecting to the World Wide Web through the current INTERNET node network, it will be possible to browse a directory of meta-information including a HEM home page giving information about its activities, e.g. HEMIS and the HEM surveys and publications.

C)   System Architecture

1.) Overview.   In previously conducted studies on HEMIS, groups of experts convened to identify the essential elements needed to construct an information system and service. The goal remains to make the system available to as much of the global community as possible, to hold information about information instead of raw data, to develop a consistent thesaurus for standardized nomenclature, to identify existing data resources and to have the system operate under DOS or Windows at the PC level.

Thus, it was determined that the system would require 3 principal software components - 1) a database management system, 2) a telecommunications and messaging system and 3) a thesaurus management system (4). Since most system designers realize the need for local, as well as, centralized data access and management, the computer industry has moved away from the traditional mainframe architecture towards a scalable, semi-open Client/Server Architecture instead. This idea still makes the system easily accessible for all users in developing and developed countries by maintaining the design concept that each client can be a standard PC. Conventional retrieval software can be used for thesauri management. An established global computer network (e.g. INTERNET or CompuServe) will be used for telecommunications and messaging. For queries concerning the documents' data, an SQL based database should make-up the back-end of the system. In other words, well established components of information technology will be linked into an integrated whole to form HEMIS.

2.) A User View of HEMIS.   When a user accesses his/her local HEMIS workstation via the standard PC, a "Visual Basic" graphical user interface (GUI) or "Windows"-like menu-driven, front-end will greet them. HEMIS will be a client/server system based on granting local access to individual PCs whose data will also be centralized in (a) designated server(s).

Sessions can generally be of two types: *data request* (traditional queries for system output) or *data suggest* (data updates for system input that will follow a quality assurance procedure).

Data Request.   From the main menu bar on the entry screen, the user will have the option to begin the session through a "Search Screen", purposefully designed to narrow down his/her search according to geographic area, dataset or catalogue, keyword (thesaurus), acronym (of the meta-database or organisation) and/or programme/activity. This will be done by typing data directly into a field or by clicking on the desired option within selection boxes.

To select the geographic area of interest, the user will manually enter the area of interest into a designated data-entry field, for example, *Munich, Germany, Europe* or *OECD countries.* For datasets, the user understands that each entry will contain thematically-organised data related to a specific topic, for example, HEM's *Survey of Environmental Monitoring & Information Management Programmes* would be seen as one dataset or entry under the general search. When selecting datasets or catalogues, the user can begin with a general search or a search by organisation. If the user prefers to begin the query by organisation, an option will permit the user to click on the organisation selection box, which will open down revealing all participating members, e.g. GRID, EEA, UNEP-HEM, etc. When the user then double-clicks on a member, for example UNEP-HEM, the associated datasets (or in HEM's case, the surveys of environmental institutes, etc.) of meta-data information will be seen as entries underneath.

If the user decides to narrow down the inquiry by keyword (see section E: Thesauri), a two-step process is required. First, the user can begin a general thesaurus hunt by keying in experimental terms (there is no possibility for incorrect data entry here because an alphabetized, corresponding term will always be returned) and then retrieving the related terms or the narrower or broader associated terms. If a language other than English is desired, an option will permit this. These multi-lingual references will still remain linked to their main keyword identifier. Next, once the desired term is attained, the user selects "Search" to automatically invoke the retrieval database engine to produce a "hitlist", or list of associated entries that match the designated query.

Data Suggest. When users wish to enter or update new data into the system, they can select an edit option from the main menu bar on the entry screen. This option will allow prospective or current participants to enter their desired data object (e.g. meta-data profile, document, etc.) into the system. If the user's data to be inserted already exists in electronic form, the system will provide several compatible, multi-platform import options. Data integrity will be maintained through a quality assurance procedure (e.g. data updates for system input will initially sit on a queue for inspection by the database administrator (DBA) and/or the quality assurance team before actually being incorporated into the "live" dataset). For each new data object, a standard system-provided entry screen will query the user for all the system-requested information. This process will, in itself, harmonize the high-level or meta-data entries.

D)   Management and Maintenance

A three-level architecture for source-data management will be implemented (6):
1.   Local - identification, collection and encoding of reference data: creating catalogues of data sources; specialized interviewers with appropriate tools, such as questionnaires, forms, manuals, thesauri, computerized software systems.
2.   Regional or national level - integration of the collected references and updating of the information system; on-line access.
3.   Central - designated organisations will integrate the various files collected by the network partners into a central data base. Archiving will also be managed at this level.

E)   Thesauri

A general multi-lingual thesaurus, based on the INFOTERRA model, will be the main tool for harmonization and standardization of nomenclature. (Initially, it is advisable to follow the EEA's model of concentrating on only three languages, with the option of expandability for future developments. In a later development phase of HEMIS, it will be desirable to include a option to search specific thesauri dedicated to more focused areas of the environmental sciences. One example of a specialized thesaurus is NASA's aerospace and engineering thesaurus.)

III.   Implementation Options

A)   Many individual organisations have developed specialized computer-aided software, e.g. UNESCO-ISIS. In some cases, standard commercial products like dBase, Fox-pro, Ingres and Word, with a customized front-end or interface have been part of a larger solution.

B)   Another option is to develop a customized client/server system unique to HEMIS's needs. This would require an on-going, iterative dialogue between the HEMIS Expert Group and the system developers during all phases of the life cycle development of the product and service. This option could aim to offer two alternatives in the future: 1) a network version of the software and 2) a subset of the network version as a standalone PC software version.

IV. Conclusions

For any international environmental system solution, thinking globally and acting locally is the challenge that the world community must face. A successful implementation scheme will focus on furnishing harmonized, accessible, information-*rich* results that suit the needs of the global community of users.

To make the HEMIS concept a reality, the HEMIS Expert Group will be called upon to recommend practical measures to move forward. Some issues to be resolved include: how to move toward actual system construction, what are the most effective and cost-efficient implementation options, what existing solutions are available, how will cooperative efforts with other organisations be possible, and what are the possible sources for funding. The purposes of HEMIS are evolving into more clearly defined, implementable terms. Thanks to the diligent efforts of partner organisations, their meta-databases and the guidance of the HEMIS Expert Group, the future for HEMIS development should move along quite rapidly given the breadth of experience from cooperating organisations.

References:

(1) Keune, H., Theisen, A., Environmental Databases and Information Management Programmes of International Organisations - Their Relevance to Environmental Management and Decision-Making Processes; Problems of Availability and Access - In: Halker, M., Jaeschke, A. (eds), Computer Science

for Environmental Protection, Informatik Fachberichte 296, 546-553, Springer-Verlag, Berlin, Heidelberg, 1991.

(2) Kampffmeyer, U.B.: Proposal for HEMIS - Design and Development Status Report, Project Consult, July 1992.

(3) An Introduction to HEM and HEM*Disk*, The Orbis Institute, May 1993.

(4) Crain, I.K., User Requirements for the Harmonization of Environmental Measurement Information System - HEMIS, March 1992.

(5) Crain, I.K., Constraints to the Extraction of Information from Complex Environmental Databases, A Discussion Paper for GTOS Workshop, 7-10 June 1994.

(6) Catalogue of Datasources for the Environment, Analysis and Suggestions for a meta-database system and service for the European Environment Agency, Joint Project EEA/Europ. Com and NRT/N. Counc.Min, Version 930831.

Adrian Manu
INFOTERM, Vienna, Austria

Terminology Standardization in the Field of the Environmental Sciences

Abstract: The paper gives an overview on pertinent standards by country and by standards organization; databases containing standardized terminology are also mentioned. Information on international terminology standardization in the field will concentrate on ISO and CEN, but also on specialized organizations such as ISWA. A project of a multilingual vocabulary on waste management, carried out by Austrian organizations in co-operation with specialized translators will be a basis for terminology standardization in Eastern European countries.

1. Introduction

Nowadays "environmental protection" has become a slogan you can read every day in the newspapers, you can hear every day on the radio and you can see every evening on TV. Everybody uses it and everybody thinks that he understands what it means. Most people think that this subject should be dealt with by the state, the authorities, etc., and does not affect them too strongly. Only in the last 2-3 years, when the amount of waste grew bigger and bigger, including household refuse, and waste sorting and separation had to be carried out by everyone, people became aware that environmental problems touch them as much as eating and drinking.

More and more countries are becoming aware of the importance of environmental protection not only for human life, but also for industrial development and progress. This evolution has certainly a strong impact on standardization in general and on terminology in particular. Almost all national standardizing bodies have already published standards in this field or are working on such documents. On the international level, terminology standardization in environmental protection is less advanced; this situation is mainly due to the fact that environmental science is a relatively young branch of science and technology, so that a unified terminology could not yet be reached. Nevertheless, efforts are being made primarily by specialized bodies having an authoritative status to quasi-standardize the respective terminology.

A more deteiled analysis of the terminology standards issued so far shows that most of them deal with waste management, waste recycling, waste dumping, handling and the transport of hazardous waste, etc. This development is quite natural since the waste industry has become one of the most important and quickly expanding branches of our economy. In general, companies specializing in this field have remained untouched by the economic crisis in Europe.

As environment and waste management are essential for human health and well-being, these fields are subject to national legislation. Standardization, therefore, has to adapt to the legal Acts and to the terminology contained in official documents. On the other hand, standards may record general usage that has already been introduces and influence in that way the language of authorities.

Technical specifications issued by big companies, testing authorities, certification bodies, etc. can also be considered as quasi-standards. They contain relevant terminology as well, partly in context, which is often stored in national databases.

2. International standardization

2.1 ISO

As mentioned before, nearly all countries of the world are aware that effective environmental

environmental protection can only be realized by international co-operation. One of the bodies where co-operation on this level has existed since the beginning of our century is the International Organization for Standardization (ISO) and its predecessor, the International Federation of National Standardizing Associations (ISA). It is, however, not easy to find common rules for such a young technology, and even more difficult to agree on a generally accepted and applied terminology. That is why work on environmental terminology has concentrated on air and water. The following standards have been published so far:

ISO 4225:1980 Air quality - General aspects - Vocabulary

It contains 60 concepts, defined in English and French. The terms are those commonly used in connection with the sampling and measurement of gases, vapours and particles for the determination of air quality. This standard is supplemented by

ISO 3649:1980 Cleaning equipment for air or other gases -

Vocabulary, which deals mainly with apparatuses and devices for the cleaning of air.

As can be seen from the year of publication, these standards are relatively old and subject to revision.

ISO 6107 Water quality - Vocabulary

This standard consists of eight parts, published between 1986 (parts 1, 5, 6) and 1993 (parts 3, 4, 8), some of them already as second editions. It is particularly comprehensive (more than 500 defined concepts), quadrilingual (English, French, Russian, German) and deals with all aspects of water technology, such as drinking water, waste water, industrial water, treatment and storage of water, sewerage, measuremenmt of water quality, water cleaning, etc.

It should be added that ISO has isued many more standards on environmental problems, most of them containing a terminology clause at the beginning, where those concepts are defined which are indispensible for the understanding of the respective standard. This is the case for nearly all standards bodies (also national ones) presented here. That means that the general number of defined concepts or terms in the field of environmental science and technology is much higher than it may seem according to this paper. The main difference between a terminology standard and a standard containig only a terminology section is that the latter is often valid only for that respective standard and cannot be generally applied .

Finally, the great importance of environment for the international community is illustrated by the fact that in 1993 was established the Technical Committee ISO/TC 207 "Environmental management", the secretariat of which is held by the Standards Council of Canada (SCC). Subcommittee 6 deals with "Terms and definitions"; its secretariat is held by the Norwegian Standards Institute (NSF). Because of its existence since one year only, the TC has not yet issued any standards.

2.2 CEN
The activity of the European Committee for Standardization (CEN) should be seen in this light: the only terminology standard published by this organization is

prEN 1085:1993 Terms and definitions in the field of wastewater treatment

Like all CEN standards, it is published in three original versions (English, French and German) and, as can be gathered form the letters "pr" before EN, it is still in a project stage. The other member states of CEN (all members of the EU and of the EFTA) have the obligation to take over the CEN standards as national standard, either in one of the three original versions or as a translation in their national language.

The draft standard prEN 1085 contains some 400 definitions, many of them based on the ISO standards mentioned above, dealing with the large field of wastewater management, one of the fields having a very strong impact on our everyday life and which will become even more important in the future, when the water resources will decrease dramatically.

As it had been said before, many of the other CEN standards dealing with environment, air, water, waste, etc., contain terminology clauses, so that the number of defined concepts is considerably higher than 400, but still by far insufficient in view of the huge subject field.

2.3 Organizations having an authoritative status
Everybody knows that the problem of waste management has become extremly pertinent today because of the immense quantities of waste and refuse polluting our environment.

The European Community is aware since longtime of this subject and also of its terminological implications. That is why the European Parliament published as early as 1984 a

Terminology of waste management

in seven languages (French, Englisch, German, Italian, Dutch, Danish and Greek). The volume contains 930 entries, each entry consisting of a term in context, without definition. The very thoroughful choice of terms makes it still usable.

The EC and the European Parliament are certainly not standardizing bodies, but as mentiond in the Introduction, some institutions and organizations have an authoritative status in a certain field or a certain geographic area, which is the case for the EC and the EP.

Another organization of the kind is the International Solid Wastes Association (ISWA). In 1992 it published (edited by John Skitt)

1000 Terms in Solid Waste Management

The languages of the vocabulary are English (with definitions), German, Spanish, French and Italian; the entries are arranged in the alphabetical order of the English terms.

3. National Standardization

As concerns national standardization, most of the countries have prepared standards on different aspects of environment, but only a few have standardized the respective terminology.

3.1 United Kingdom
The British Standards Institution (BSI) for instance, has taken over the above mentioned ISO Standard 6107 and has published it under the number

BS 6068 Sections 1.1 to 1.8 Water quality. Glossary

Some parts of this standard are still based on first editions of the respective parts of the ISO standard. Another British standard, also based on an ISO standard is

BS 6069:Part 1:1981 Methods for characterization of air quality. Part 1. Units of measurement

3.2 France
France is one of the countries which began very early to standardize terminology in the different subfields of environment. The first standard of this kind was published in 1972 by the French standards body the Association francaise de normalisation (AFNOR):

NF X 30-001, dec. 1972 Environnement - Définition des termes généraux [Environment, Definition of general terms]

Only some 30 concepts of a very general nature are defined; English equivalents are given.
A nonterminological standard which contains a large terminology section is

X 30-200, apr. 1993 Système de management environnemental [System of environmental management]

Contains concepts for modern management of the environment.

X 31-071, feb. 1983 Qualité des sols - Matériaux types - Définitions - Prélèvement [Soil quality - Standard materials - Definitions - Sampling]

Biotechnology can certainly also be considered as an aspect of environmental engineering. AFNOR published the series X 42 which deals with this subject:

NF X 42-000, dec. 1986 Biotechnologies - Vocabulaire - Termes généraux [Biotechnology - Vocabulary - General terms]

NF X 42-001, dec. 1986 Biotechnologies - Vocabulaire - Génie enzymatique [Biotechnology - Vocabulary - Enzyme engineering]

NF X 42-002, dec. 1987 Biotechnologies - Vocabulaire - Génie immunologique [Biotechnology - Vocabulary - Immunological engineering]

NF X 42-003, dec. 1988 Biotechnologies - Vocabulaire - Plantes transgéniques et cultures cellulaires [Biotechnology - Vocabulary - Transgenic plants and cell cultures]

X 42-004, jul. 1988 Biotechnologies - Vocabulaire - Génie génétique [Biotechnology - Vocabulary - Genetic engineering]

NF X 42-005, aug. 1988 Biotechnologies - Vocabulaire - Filtration, microfiltration, ultrafiltration, osmose inverse [Biotechnology - Vocabulary - Filtration, microfiltration, ultrafiltration, reverse osmotics]

AFNOR dealt as well quite early with air quality and published the following standards:

NF X 43-001, aug. 1982 Qualité de l'air - Vocabulaire [Air quality - Vocabulary]

The standard covers 66 concepts of a very general nature. For each French term an English equivalent is given.

NF X 44-001, feb. 1981 Séparateurs aérauliques - Vocabulaire [Air separators - Vocabulary]

Some 60 concepts deal with apparatus, but also with other technical and physical aspects of air cleaning. The French terms are accompanied by English equivalents.

NF X 44-102 Enceintes à empoussièrement controlé - Définitions - Classification [Controlled dust content chambers - Definitions - Classification]

The terminology section of this standard gives the basic concepts for the field mentioned in the title.

3.3 Germany

Germany is one of the most advanced countries in the world as regards environmental protection. Some 800 standards, technical rules, codes of practice, laws, decrees, etc. control all aspects of environmental activities carried out either by authorities, by companies or by private persons. All these documents certainly contain the respective terminology in context, many of them also in the form of a terminology clause. As the present paper considers only standards, it should be mentioned that not only the German Institute for Standardization (DIN) publishes German standards, but also the Association of German Engineers (VDI) and the Association of German Electrical Engineers (VDE). In the following, the most important German terminology standards are presented.

DIN 4045, dec. 1985 Abwassertechnik - Begriffe [Waste water engineering - Concepts]

A very comprehensive document containing nearly 500 German terms with definitions and English equivalents, arranged in systematical order; alphabetical indexes for search purposes are placed at the beginning of the standard.

DIN 4046, sep. 1983 Wasserversorgung - Begriffe [Water supply - Concepts]

This standard constitutes at the same time technical regulations of the DVGW (German Association for Gas and Water). It contains some 200 terms with definitions, some of them accompanied by symbols, formulas and illustrations, covering the whole field of water supply engineering.

DIN 4047, jan. 1993. Landwirtschaftlicher Wasserbau - Allgemeine Begriffe [Water enginering of agricultural lands - General concepts].

The document lists the basic concepts of the field mentioned in the title, under special consideration of ecological aspects. English equivalents are given for the German terms.

DIN 30702-5, oct. 1987 Kommunalfahrzeuge - Begriffe für Saugfahrzeuge und Hochdruck-Spülfahrzeuge [Municipal vehicles - Concepts for suction vehicles and high pressure flushers]

One page of basic concepts for street cleaning vehicles .

DIN 30706-1, mai 1991 Entsorgungstechnik - Begriffe f r Hausabfallentsorgung und Entsorgungsfahrzeuge [Waste disposal technology - Concepts for household waste disposal and waste disposal vehicles]

Defines the different kinds and categories of waste as well as the possibilities of disposal and dumping.

DIN 30706-2, nov. 1985 Begriffe der kommunalen Technik - Straßenreinigung [Concepts for communal purposes - Street cleaning]

23 definitions on waste, litter, refuse and the possibilities for collecting them.

DIN ISO 4225, mai 1986 Luftbeschaffenheit - Allgemeine Gesichtspunkte - Begriffe [Air quality - General aspects - Vocabulary]

German translation of the ISO standard described under item 2.1.

DIN ISO 6879, jan. 1984 Luftbeschaffenheit - Verfahrensgrößen und verwandte Begriffe für Messverfahren zur Messung der Luftbeschaffenheit [Air quality - performance characters and related concepts for air quality measuring methods]

VDI 3490-1, dec. 1980 Messen von Gasen - Prüfgase - Begriffe und Erläuterungen [Measurement of gases - Calibration gas mixtures - Concepts and explanations]

VDI 3491-1, sep. 1980 Messen von Partikeln - Kennzeichnung von Partikel - dispersionen in Gasen - Begriffe und Definitionen [Particulate matter measurement - Characteristics of suspended particulate matter in gases - Terms and definitions].

3.4 Austria

Austria also started very early to control environmental pollution. Like in Germany, there are a great number of ministerial decrees, orders, technical rules, laws, etc. for environmental purposes.

In standardization, accent has been placed on economic and ecological energy usage on the one hand, and on waste management on the other hand.

ÖNORM M 7101 to 7160 Grundbegriffe der Energiewirtschaft [Basic concepts of energy economy]

has been published between 1984 and 1991. The series of 16 standards covers all aspects of energy, beginning from the sources of energy to economy of energy, including methods and rules for environmental protection.

ÖNORM S 2000 to 2008 Abfall - Begriffe [Waste - Concepts]

was published between 1977 and 1992. The series deals with generalities, collection and transport of wastes, methods of processing, treatment recycling and deposit of wastes as well as with zhe management of contaminated sites.

ÖNORM S 2100, mar. 1990 Abfallkatalog [Catalog for waste] lists all known categories of wastes.

ÖNORM S 2100, jun. 1993 Katalog gefährlicher Abfälle [Catalog of hazardous waste]

lists all known categories of hazardous wastes.

In 1992 the Austrian Standards Institute, in collaboration with the International Information Centre for Terminology (Infoterm) and the International Network for Terminology (TermNet) started a pilot project financed by the Austrian Federal Ministry for Environment, Youth and Family concerning the establishment of a nine language vocabulary of waste management. 500 standardized concepts on waste were selected from the terminology databank run by Infoterm for the Austrian Standards Institute, for which equivalents in English, Russian, Czech, Slovak, Polish, Croatioan, Slovenian and Hungarian have been added (including definitions) by specialized mother-tongue translators, and finally checked by subject specialists from the respective countries. The result of this work will be published by the Ministry and by the Austrian Standards Institute in the fall of 1994.

This project is part of an Austrian assistance program for Eastern European countries and will be the basis for standardization of waste terminology in these countries. Follow-ups of this project are planned, e.g., addition of other languages, extension of the scope, etc.

These examples represent only a selection from the very comprehensive field of national, regional and international standardization of environmental terminology. Many more countries are standardizing their environmental terminology, others have not yet begun. The paper presented might provide an additional impetus for these latter ones to start this indispensable work as well as to provide models based on the examples given.

List of Authors

Agnes Ajtay, Eötvös Loránd University, Department of Cartography,
Ludovika tèr2, H-1083 Budapest, Hungary

Wolf-Dieter Batschi, Umweltbundesamt, Bismarkplatz 1, D-14193 Berlin

Hassane Bendahmane, INFOTERRA PAC, UNEP, P.O.Box 30552, Nairobi, Kenya

Heiner Benking, FAW Research Institute for Applied Knowledge Processing at
Ulm University, P.O.Box 2060, D-89010, UlmBENKING@faw.uni-ulm.de,

I.Beseda, Fac. of Ecology, State Forest Products Research Institute, Bratislava,
Slovakia

Dr.Gerhard Budin, IITF, Sensengasse 8, A-1090 Wien

Ewa Chmielewska-Gorczyca, Institute for Computer and Information Engineering
ul. Spasowskiego 13 m.18, PL-00-348 Warszawa, Poland

Dr. Ingetraut Dahlberg, International Society of Knowledge Organization
36a Woogstrasse, D-60431 Frankfurt

Dr.Maria Domokos, Korny.Ved.Felugy, Rakoczi ut 41, H-1088 Budapest, Hungary

Dr.Endre Dudich, Karolyi M. u 14B IV.51, H-1053 Budapest, Hungary

Ivan Duša, Ministry of the Environment of the Slovak Republik
Hlboká 2, 81235 Bratislava, Slovakia

Dr.Bruno Felluga, Reparto Ricerca e Documentatione Ambientale,
Istituto Technologie Biomediche, Via Morgani 30/E, I-00161, Italy

Peter M.Frischknecht, ETH Zürich, Department of Environmental Sciences,
Voltastr. 65. CH-8092 Zürich, Switzerland

Ch.Galinski, INFOTERM, Heinestrasse 38, A-1021 Wien, Austria

Jan Gotthard, CEIT, Biskupicka 1, 821 06 Bratislava, Slovakia

Anthony J.N. Judge, Union of International Associations
40 rue Washington, B-1050 Brussels, Belgium

Vladimir Kašša, National Centre for Informatics,
Hanulova 5/a, 84416 Bratislava, Slovakia

S. Katuščak, Faculty of Ecology, Technical University
Fadruszova 3, 84105 Bratislava, Slovakia

Dr. Hartmut Keune, UNEP-HEM c/o GSF Neuherberg, PF 1129
D-85758 Oberschleissheim

Zdravko Krakar, Institute for Information Technologies,
Mazuraniev trg 8, 41000 Zagreb, Croatia

Tobias Krull, Institute of Veritology, Mittelstr. 19, D-32108 Bad Salzuflen

Dr. Hellmut Löckenhoff, Ossietskystr 14, D-71522 Backnang

Sandra Lucke, Reparto Ricerca e Doc.Ambientale, Ist. Tecnol.Biomediche, CNR
Via Morgagni 30/E, I-00161 Roma

Adrian Manu, INFOTERM, Heinestr.38, A-1021 Wien, Austria

J.L.McKenna, UNEP-HEM c/o GSF, Neuherberg,
PF 1129, D-857758 Oberschleißheim

208

Wim W.de Mes, Unesco/IHP, Ch.de Bourbonstraat 6,
NL-1732 KJ Noord Scharwoude

Nenad Mikulić, Ministry of Civil Engineering and Environmental Protection,
Vukovarska 78, 41000 Zagreb, Croatia

Csaba Molnar, Eötvös Loránd University, Department of Cartography
Ludovika tér 2, H-1083 Budapest, Hungary

Dr.M.Muraszkiewicz, Institute for Computer and Information Engineering
Lokajskiego 16 m 22, PL-02793 Warszawa, Poland

Prof.Dr.O. Nacke, Institut für Veritologie, Mittelstr. 19, D-32108 Bad Salzuflen

Jasna Novak, Ministry of Justice, Agency for Informatization, Savak Cesta 41,
41000 Zagreb, Croatia

Maria Palmera, Reparto Ricerca e Doc.Ambientale, Ist. Tecnol.Biomed., CNR,
Via Morgagni 30/E, I-00161 Roma

J.Pajtik, Faculty of Ecology, Technical University, Zvolen, Slovakia

Dr.Werner Pillmann, CEDAR, Marxergasse 3/20, A-1030 Wien, Austria

Ulla Pinborg, National Forest and Nature Agency
Haraldsgade 53, DK-2100 Copenhagen Ö, Denmark

Paolo Plini, Reparto Ricerca e Doc. Ambientale, Ist. Tecnol.Biomed., CNR
Via Morgagni 30/E, I-00161 Roma

Dr. W.-F. Riekert, Forschungsinstitut für anwendungsorientierte Wissensverar-
beitung (FAW) an der Univ. Ulm, PF 2060, D-89010, Ulm
riekert@faw.uni-ulm.de

H.Rybinski, ICIE, Lokajskiego 16 m 22, Warsaw, Poland

Prof.Dr.Klaus-Dirk Schmitz, Fachschule Köln, Fachbereich Sprachen, Mainzer Str.
5, D-50678 Köln

Prof.Dr.Roland W. Scholz, Department of Environmental Sciences, Swiss Federal
Institute of Technology Zürich (ETHZ), Voltastr. 65, CH-8092 Zurich
Switzerland

Krystiyna Siwek, Information Processing Centre
Al. Niepodleglosci 188b, PL-00950, Warszawa, Poland

Dr.Pavla Stančíková, CEIT-Centre of ECO-Information & Terminology
Biskupichá 1, 82106 Bratislava, Slovakia

Jela Steinerová, Department of Library & Information Science, Faculty of Arts,
Comenius University, Gondova 2, 81801 Bratislava, Slovakia

Eva Stengård, Swedish Museum of Natural History, Documentation Center
Box 50007, SE-10405 Stockholm, Sweden

Dr.Elmar A. Stuhler, Technische Universität München, Institut für Wirtschafts-
und Sozialwissenschaften, D-85350 Freising

Tamás Toth, Eötvös Loránd University, Department of Cartography
Ludovika tér 2, H-1083 Budapest, Hungary

Dr. Irene Wormell, Royal School of Librarianship, Birketinget 6,
DK-2300 Copenhagen S, Denmark

Dr. Konrad Zirm, Bundesministerium für Umwelt, Jugend und Familie
Sektion 1, Radetzkystr. 2, A-1031 Wien, Austria

Name and Subject Index